Cambridge Elements ≡

Elements in Metaphysics
edited by
Tuomas E. Tahko
University of Bristol

REDUCTION, EMERGENCE, AND THE METAPHYSICS IN SCIENCE

Exploring New Foundations

Carl Gillett
Northern Illinois University

CAMBRIDGE
UNIVERSITY PRESS

CAMBRIDGE
UNIVERSITY PRESS

Shaftesbury Road, Cambridge CB2 8EA, United Kingdom

One Liberty Plaza, 20th Floor, New York, NY 10006, USA

477 Williamstown Road, Port Melbourne, VIC 3207, Australia

314–321, 3rd Floor, Plot 3, Splendor Forum, Jasola District Centre,
New Delhi – 110025, India

103 Penang Road, #05–06/07, Visioncrest Commercial, Singapore 238467

Cambridge University Press is part of Cambridge University Press & Assessment,
a department of the University of Cambridge.

We share the University's mission to contribute to society through the pursuit
of education, learning and research at the highest international levels of excellence.

www.cambridge.org
Information on this title: www.cambridge.org/9781009500951

DOI: 10.1017/9781009083423

First published 2025

A catalogue record for this publication is available from the British Library

ISBN 978-1-009-50095-1 Hardback
ISBN 978-1-009-08759-9 Paperback
ISSN 2633-9862 (online)
ISSN 2633-9854 (print)

Reduction, Emergence, and the Metaphysics in Science

Exploring New Foundations

Elements in Metaphysics

DOI: 10.1017/9781009083423
First published online: January 2025

Carl Gillett
Northern Illinois University
Author for correspondence: Carl Gillett, carl.gillett@gmail.com

Abstract: This Element offers a fresh treatment of the two cycles of reduction–emergence debates in the sciences and their "reductionist" and "emergentist" positions. It suggests philosophers have neglected the compositional models/explanations, and the "endogenous" kind of metaphysics, central to these debates. It highlights how such endogenous metaphysics underpins what is termed the Dynamic Cycle, by which scientists develop novel ontological concepts to underwrite new models/explanations to solve scientific problems. And it subsequently shows that the reductionist and emergentist views in the scientific debates follow the Dynamic Cycle. In the first cycle of debates, in the early twentieth century, the Element outlines how "Everyday Reductionism" pioneered a novel family of compositional models/ explanations in one of the most successful research movements in twentieth-century science. And, in current debates, it frames contemporary emergentist positions offering ontological innovations, underwriting new families of models, to address problems at the cutting edge of twenty-first century science.

Keywords: reductionism, emergentism, scientific metaphysics, scientific method, emergence

ISBNs: 9781009500951 (HB), 9781009087599 (PB), 9781009083423 (OC)
ISSNs: 2633-9862 (online), 2633-9854 (print)

Contents

Introduction

Over the last century, working scientists have twice been gripped by reduction–emergence debates that they took to have a direct, and significant, impact on their research. These scientific discussions have revolved around apparently metaphysical issues about the structure of nature such as the extent of compositional relations and models/explanations, the existence of downward whole-to-part determination, and the character of the fundamental laws, among others. The centrality of compositional models/explanations to these debates, for example, is seen in the famous slogans of the rival views in these discussions. Various kinds of reductionists in the sciences have famously claimed "Wholes are nothing but their parts," though they often mean very different things by this in distinct periods, while emergentists have contended that "Wholes are more than the sum of their parts" and, more recently, that "Parts behave differently in wholes."

Unfortunately, philosophical discussions of reductionism have overlooked these scientific debates and positions. Instead, discussions in the philosophy of science have famously focused on the Nagelian or semantic framework for "reduction" concocted by positivist philosophers of science (Nagel 1961). The positivists programmatically dismissed metaphysics as literal nonsense, hence mapping a very different course than scientific debates. Under the Nagelian view, the products of science are "theories" in big groups of statements including law statements; scientific explanation is derivation from law statements in the so-called deductive-nomological view (Hempel 1965); and "reduction" is hence the derivation, and putative explanation, of the law statements of one theory from the laws of another theory using identity statements. We thus have a purely semantic view of "reduction" and "reductionism" focusing on statements and putatively emptied of any ontology or metaphysics.

The initial problem loudly trumpeted for this programmatic philosophical picture of reduction came from writers like Fodor (1974) whose Multiple Realization Argument showed that in the higher sciences we do not get the identities required for Nagelian reduction.[1] But further problems have subsequently been more quietly established. First, philosophers of science have shown the main products of the higher sciences are models rather than such "theories." Second, it has been noted that the higher sciences, such as biology, offer few laws. Third, philosophers of science have accepted that explanation is not derivational and hence that the deductive-nomological view is badly mistaken. And fourth, writers have outlined how integration between models, rather

---

[1] Kitcher (1984) also offered the Predicate Indispensability Argument showing that the predicates of the higher sciences are not dispensable in the ways Nagelian reduction entails.

than derivational relations between "theories," is what we in fact find in the higher sciences (Brigandt 2011). Overall, the consensus philosophical conclusion is that Nagelian reduction – and hence "reductionism" – is absent from scientific practice in the higher sciences.

The latter problems do indeed make a compelling case for the consensus view that the *philosopher's* notions of "reduction" and "reductionism" have little application to the higher sciences. But what we still lack is any account of what *scientists* have been concerned about in their voluminous discussions of "reductionism," let alone "emergentism," for the past 150 years.[2] My goal here is therefore to supply our missing treatments of the sets of reduction–emergence debates in the sciences themselves and the main "reductionist" and "emergentist" positions that they involve. My secondary aim is to illustrate the overlooked scientific practices, involving what I term endogenous metaphysics, driving the latter discussions that I show are central in the sciences.[3]

As well as the unfortunate focus on Nagelian reductionism, I suggest we can also trace the philosophical neglect of the scientific reduction–emergence debates to two other, connected sources. First, philosophers of science, and philosophers more widely, still do not recognize the family of compositional models/explanations that are common in many sciences and, albeit in different ways, to the two sets of scientific debates. And second, many philosophers of science continue to take metaphysics to be something solely done by analytic philosophers whose practices they take to be unproductive for scientific research.

Given this background, I therefore begin my work, in Section 1, by seeking to reset our view of science in some foundational respects. Looking at a concrete case in contemporary physiology, cell biology, and molecular biology, I show that researchers in fact give a far wider variety of ontic models/explanations beyond just causal or mechanistic ones, including a family of philosophically overlooked compositional models/explanations. Furthermore, I highlight how these models are in plural, but integrated groups I term coalitions.

Once we better appreciate the plural array of ontological concepts and models/explanations that scientists use, then we are immediately hit with

---

[2] See Brigandt and Love (2023) for a summary of the philosophical consensus – one lacking scientific conceptions of "reductionism" built around compositional models/explanations. And see Mitchell (2009) for a thorough critique of Nagelian reduction as adequate to contemporary higher sciences.

[3] I therefore put to one side here the philosophical debates over "reduction" and "emergence" to focus on better understanding the scientific debates. But see Gillett (2002a) for a survey of philosophical "emergence" debates, and Aizawa and Gillett (2014) for a survey of philosophical discussions over "reduction." I have also previously documented at book length (Gillett 2016a) how these philosophical accounts of reductionism and emergentism are often detached from the scientific debates.

some new questions that do not arise if there is only one kind of model/ explanation. One such question is *how* do researchers develop new kinds of model/explanation? I sketch an abstract general answer in what I term the Dynamic Cycle of ontic concept/model/explanation creation, application, assessment, and revision. I suggest the Dynamic Cycle spans both the "context of understanding," where scientists develop such concepts and models/ explanations to describe, explain, and understand nature, as well as the "context of investigation," where the resulting models are applied by researchers, along with other models, to generate empirical findings. I highlight how scientists plausibly assess the ontological concepts/models developed in the context of understanding using the resulting empirical findings and hence decide whether to supplement, revise, or replace them – resulting in the Dynamic Cycle being iterative and potentially progressive.

The Dynamic Cycle usually focuses on *local*, incremental, ontological innovations by scientists in their existing models and hence extant categories of entity. But I also note how appreciating the Dynamic Cycle opens the possibility that scientists might sometimes innovate through *global* and/or *categorial* ontological innovations, namely the development of whole new ontological categories, to address long-standing scientific problems by developing new families of model. I speculate that such broader innovations would be pursued, if they exist, through groupings of like-minded scientists that I term Global Ontological Research Movements. I suggest that one obvious question is whether the positions pressed by researchers in scientific reduction–emergence debates are such movements.

Drawing together the work of Section 1, I conclude that what I dub endogenous metaphysics is central to key scientific practices such as the Dynamic Cycle's development of novel ontological concepts to underwrite new models/explanations to help with ongoing scientific problems. I note the sharply contrasting features of this endogenous metaphysics and what I term the exogenous metaphysics that philosophers of science take to be practiced in analytic philosophy. Unfortunately, I flag how philosophers of science have plausibly taken the exogenous metaphysics they ascribe to philosophers as their *sole exemplar* of metaphysics. I suggest that many philosophers of science have consequently overlooked, or explicitly dismissed, a place for endogenous metaphysics in scientific practice.

This is obviously a particular problem for philosophical treatments of the two cycles of reduction–emergence debates if they are driven by Global Ontological Research Movements. I therefore turn to providing specific stories about how particular kinds of model/explanation have been developed in tandem with research movements advocating ontological innovations that underwrite these models. In Section 2, I return to the first cycle of reduction–emergence debates that began in the late nineteenth and early twentieth centuries. This cycle of debates focused on seemingly intractable scientific problems. For instance, in chemistry, properties of substances, like common salt, had resisted explanation using their lower-level constituents, like sodium and chlorine, while in biology, for example, the digestive activities of the stomach also had persistently resisted explanation using its lower-level constituents, whether cells or molecules.

I show that we can plausibly interpret the two main positions in these early reduction–emergence debates as both being Global Ontological Research Movements focused on addressing these problems using the Dynamic Cycle. On one side, I outline how we have what I term the Ontological emergentism of so-called Vitalists and their allies that presses the ontological innovation of taking some activities and properties of chemical and biological individuals to be uncomposed and to involve "special" uncomposed kinds of force and/or energy. On the other side of the debates, and more importantly given the subsequent history, we have what I term Everyday Reductionism, whose onto-logical innovation was its claim that not only higher-level individuals, but also all their activities and properties, are fully composed by lower-level parts and their activities and properties.

Given its subsequent importance, I focus on Everyday Reductionism whose ontological innovation underwrites a whole new family of ontic models/explanations in the philosophically neglected suite of systematically integrated compositional models/explanations surveyed in Section 1. Crucially, I detail how the new models/explanations provided by Everyday Reductionism allow scientists not only to address the long-standing problem cases but also to make productive applications of such models in a wider range of sciences, including psychology and the neurosciences.

I conclude that Everyday Reductionism is plausibly just the kind of Global Ontological Research Movement I predicted might sometimes arise in the sciences if researchers use the Dynamic Cycle. And I note that Everyday Reductionism has distinctive features including (i) a global ontological innov-ation which underpins a new guiding picture of nature; (ii) a new family of ontic models/explanations, underwritten by (i); and (iii) new methodologies driven by (i) and (ii).

Appreciating Everyday Reductionism begins to illuminate the story of how compositional models/explanations were developed in various sciences. For I briefly highlight how Everyday Reductionism was one of the most famous, and successful, research movements of twentieth-century science. In fact, by the middle of the twentieth century, the first set of reduction–emergence debates were taken to be *resolved* in the sciences by the work of the Everyday

Reductionist movement. Successful compositional explanations had not only been provided for the starting problem cases but were also supplied in a huge array of other sciences. The familiar headline cases include how the inheritance of biological traits is compositionally explained by molecules like DNA and RNA, but quotidian work included successfully supplying compositional models/explanations in manifold cases across various sciences.

Today everyone in the sciences is thus plausibly a reductionist in the sense of accepting Everyday Reductionism and endorsing the ubiquity and utility of compositional models/explanations, and hence many levels of entities and sciences. And I note how both sides in our second, contemporary set of reduction–emergence debates – self-proclaimed reductionists and emergentists alike – each endorse Everyday Reductionism.

In the final sections of this Element, I explore whether our *present* cycle of reduction–emergence debates also involve Global Research Ontological Movements pursuing the Dynamic Cycle. I start, in Section 3, by looking at the *other* reductionism in the sciences that I suggest both philosophers and scientists routinely conflate with Everyday Reductionism. This is what I follow Nancy Cartwright (1994) in terming Fundamentalism, though I differ with Cartwright in my understanding of this view, which I argue takes Everyday Reductionism as both its starting point and its target. For I highlight how Fundamentalism is driven by theoretical arguments, in ontological parsimony reasoning, focused on Everyday Reductionism's vast array of successful compositional models/explanations. The Fundamentalist uses such parsimony reasoning to argue that *if* we have successful compositional models/explanations in some case, *then* we should conclude that *we have parts alone*, rather than any wholes in this example – hence subtracting much of the guiding picture of nature pressed by Everyday Reductionism.

Fundamentalism instead offers a guiding picture of nature under which there are no compositional levels of parts and wholes, but only organized, interrelated collectives of parts of varying scales. I also bring out how Fundamentalism implicitly endorses an ontological assumption in what I term the Simple view of nature under which the individuals that are parts always behave in the same ways, and hence are covered by the same determinative laws or principles, under all conditions. For example, I show that the Simple view of nature is a precondition of the truth of the Final Theory that Fundamentalists like Steven Weinberg (1992) and E. O. Wilson (1998) dream about.

Overall, I conclude that Fundamentalism aspires to be a Global Ontological Research Movement, since it mirrors Everyday Reductionism in feature (i) by pressing a global ontological innovation and a new guiding picture of nature built upon it; and also in characteristic (iii) by providing new methodologies deriving from (i).

But I suggest we should place an asterisk by this claim. For I note that Fundamentalism does not offer researchers any new models/explanations to use to address their ongoing problems, hence it lacks feature (ii). Instead, Fundamentalism focuses on using theoretical arguments to explore the import of our extant models, in compositional models/explanations, rather than seeking new resources to address ongoing problems. Unlike Everyday Reductionism, Fundamentalism thus does not pursue the Dynamic Cycle of offering onto-logical innovations to underwrite novel ontic models/explanations to address ongoing scientific difficulties.

To see whether other positions in our present reduction–emergence debates do pursue the Dynamic Cycle, in Section 4 I look at an exemplar kind of problem in contemporary science in what I term Challenging Compositional Cases. From the behavior of electrons in superconductors, to the activities of proteins in cells, I highlight how our present knowledge in such cases points to a failure to understand, and explain, the behavior of the relevant parts despite our having successful compositional and other models/explanations – hence framing an ongoing challenge in various sciences.

I then outline how writers in science and philosophy have sought to address Challenging Compositional Cases using the Dynamic Cycle and a couple of global ontological innovations. To start, I highlight the innov-ation I term the Conditioned view of nature inspired by the findings from Challenging Compositional Cases. The Conditioned view claims that indi-viduals sometimes act differently under certain conditions than they would behave if the laws or principles in simpler systems exhaustively applied. Under the Conditioned view of nature, parts may thus have what I term "differential" activities and powers – that is, activities and powers that are different from those the parts would have if the laws or principles in simpler systems were exhaustive.

Most significantly, I highlight how adopting the Conditioned view, either alone or in combination with another ontological innovation, underwrites *two* distinct positions each offering a new family of models/explanations that provides fresh resources with the ongoing problems in Challenging Compositional Cases. I examine the simplest of these positions in the remainder of Section 4 in what I term the Causally Conditioned view. This position accepts *both* the Conditioned view and embraces differential powers/activities of certain parts *and* claims that the differential activities of these parts are causally, or at least diachronically, determined by other parts or individuals at the same level. I dub the resulting models Causally Conditioned models/explanations and I highlight how these new models/explanations offer potential help in Challenging Compositional Cases.

In Section 5, I then turn to the other position in the most prominent kind of contemporary scientific emergentism or what I term Mutualism. This Mutualist position embraces *two* global ontological innovations. First, such emergentists also embrace the Conditioned view of nature and accept "Parts behave differently in wholes" because parts have differential powers/activities in certain complex wholes. But second, Mutualists endorse a further ontological innovation in a novel kind of whole-to-part determination, what I term machretic determination, by which wholes determine the differential powers/activities of these parts.

Putting these two innovations to work together, I sketch how Mutualism offers a new guiding picture of nature under which wholes and parts are *mutually determinative* and interdependent, hence the name for the view, where "Wholes are more than the sum of their parts" while still being comprehensively composed. I also carefully note how the Mutualist's guiding picture of nature *supplements* the picture of Everyday Reductionism, rather than subtracting from it as Fundamentalism advocates.

Given this pair of ontological innovations, I highlight how Mutualism underwrites a new family of what I term Mutualist models/explanations, again offering novel resources in Challenging Compositional Cases. And I mark how these novel Mutualist models/explanations *supplement* the compositional models/explanations of Everyday Reductionism and other models in our present coalitions of models.

I thus conclude that the Mutualism of contemporary scientific emergentism is also a Global Ontological Research Movement fully mirroring all three characteristic features of Everyday Reductionism. Mutualism has (i) novel global and categorial ontological innovations and a new guiding picture of nature built upon them; (ii) a new family of models and explanations underwritten by its novel ontological innovations; and (iii) new methodologies underpinned by (i) and (ii). Somewhat surprisingly, given the terminology we have fallen into, I therefore suggest it is emergentist Mutualism that is the real intellectual heir to Everyday Reductionism in contemporary science, rather than Fundamentalist reductionism.

By the end of this Element, in light of these findings, I hope you will better appreciate the nature of the endogenous metaphysics that we see in the sciences and that you will also understand why we need to reset our narratives about reductionism and emergentism in actual scientific practice. Endogenous metaphysics is not like the exogenous metaphysics pursued by philosophers, but I hope you will start to see how, and why, it continues to be a highly productive endeavor in the sciences. Furthermore, finally understanding the Everyday Reductionism of working scientists allows us to appreciate one of the most successful research programs of twentieth-century science and to understand how our ubiquitous compositional models/explanations were developed. And, turning to today's science, I hope you will glimpse the exciting positions that continue to offer ontological innovations, in the service of the Dynamic Cycle, to address problems at the cutting edge of contemporary science.

1 Revisiting the Variety, Nature, and Development of Ontic Models/Explanations

Dominant views in philosophy of science have held that there is one kind of scientific explanation. Initially, the positivist's view that "All explanation is derivational" held sway, but more recently "All explanation is causal" and "All explanation is mechanistic" have been popular. If there is only one kind of unchanging thing that is a scientific explanation, then there is, of course, no pressing question about how, further, new kinds of model and hence explanation come into existence, or whether broad ontological innovations might underwrite such novel families of models/explanations. On the contrary, however, if there are various kinds of model/explanation in the sciences, then it becomes a real issue how researchers develop such new kinds of model/explanation and how this relates to the development of novel ontological conceptions.

My goals in this section are, therefore, as follows: in Section 1.1, I begin to highlight how contemporary science does in fact use a plural array of distinct, but integrated, ontic models/explanations. I highlight how this includes a philosophically overlooked family of compositional models/explanations that I briefly explore in Section 1.2. I also highlight, in Section 1.3, the larger groups of models offered in the context of understanding.

Having found such a plural array of models, in Section 1.4, I therefore sketch an abstract general answer about how new ontic concepts and models/explanations are developed centered around the Dynamic Cycle depicted in Figure 1. I also highlight how there is space for applications of the Dynamic Cycle utilizing categorial or global ontological innovations to underwrite whole new families of ontic models/explanations. And I raise the possibility that "reductionist" and "emergentist" positions in the two cycles of scientific reduction–emergence debates are just these kinds of movement. Lastly, in Section 1.5, I begin to illuminate the different kind of metaphysics that drives such innovative work in the sciences in what I am terming endogenous metaphysics.

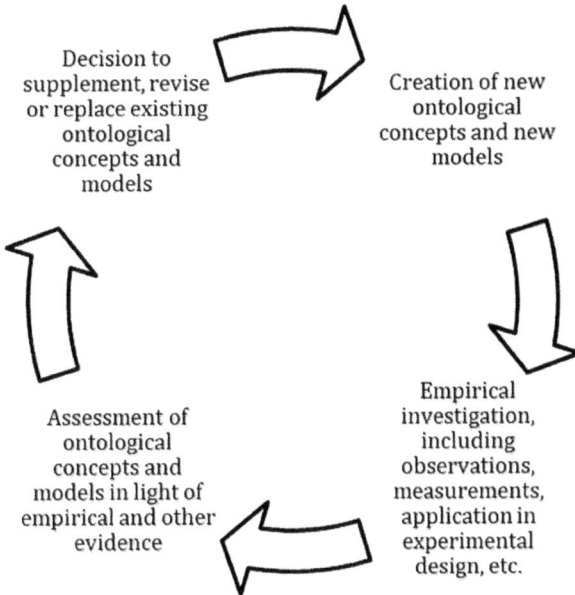

Figure 1 Abstract diagram of the iterative "Dynamic" cycle of creation, empirical application, assessment, and alteration of ontic models/explanations and their underlying ontological concepts.

1.1 Compositional and other Ontic Models/Explanations in the Sciences

I look at a mature, concrete case from physiology, cell biology, and molecular biology in our accounts of skeletal muscles and their contraction, built around the sliding filaments model and others, framed in Figure 2. Looking at this case allows us to explore the models/explanations that researchers give in the context of understanding when they are seeking to provide descriptions, explanations, and understanding of nature. In Section 1.1.1, I look at some of the types of intralevel models/explanations used in this example. Then, in Section 1.1.2, I consider some of the kinds of interlevel models/explanations. In Section 1.1.3, I outline why we have found more kinds of ontic models/explanations than just causal or mechanistic ones and I summarize the varieties we have surveyed.

1.1.1 Some Intralevel Models/Explanations

To start, let us mark that in response to the question "Why did that bone move?" one good answer, at the organ level, is "Because the connected skeletal muscles contracted." As Figure 3 highlights, in this explanation we use a model backed by activities of muscles, their contracting, to explain the

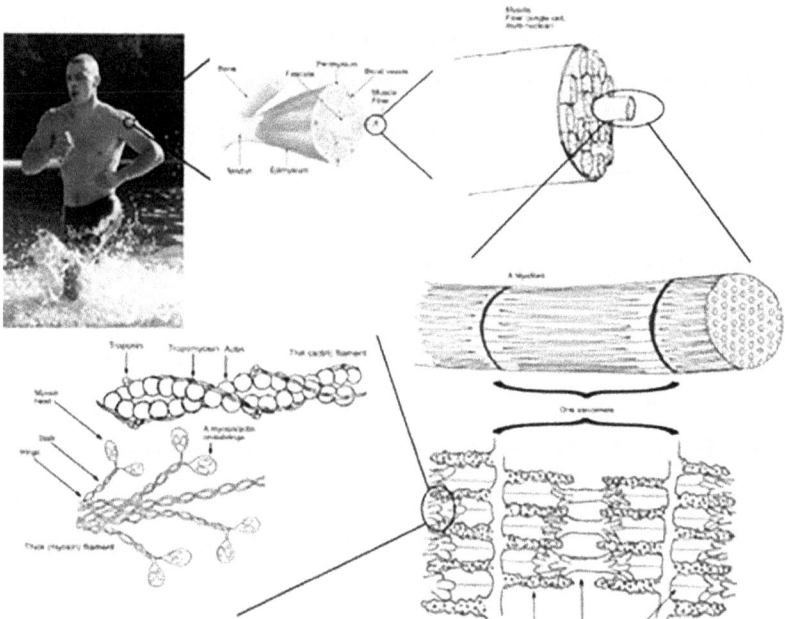

Figure 2 The basis of the famous sliding filaments model of skeletal muscle contraction. (Wikimedia commons image created by Raul654 distributed under CC-BY 3.0 license: https://en.wikipedia.org/wiki/Muscle_contraction#/media/ File:Skeletal_muscle.jpg)

bone's movement in what is termed an etiological mechanistic explanation by the New Mechanists in philosophy of science.[4] This is a type of causal explanation where we explain some effect using thick causal relations in activities of various individuals.[5]

Alongside such models/explanations, though neglected by philosophers, researchers routinely offer explanations of such activities using properties of the relevant individual. For instance, in response to the question "Why did the skeletal muscle contract with that force?" one good answer, in the appropriate context, is "Because the skeletal muscle has a certain strength." In the model underlying this explanation, an individual, here a skeletal muscle, is represented as instantiating a certain property that, under appropriate conditions, results in an activity of that individual, here contracting, and hence explains this activity.

---

[4] See, for example, New Mechanists such as Craver (2007), among others.
[5] The reader should note that I am a pluralist about the concepts of causation used in the sciences. I take "thick" causal concepts to include relations of activity and "thin" causal concepts to be captured by manipulability or difference-making accounts.

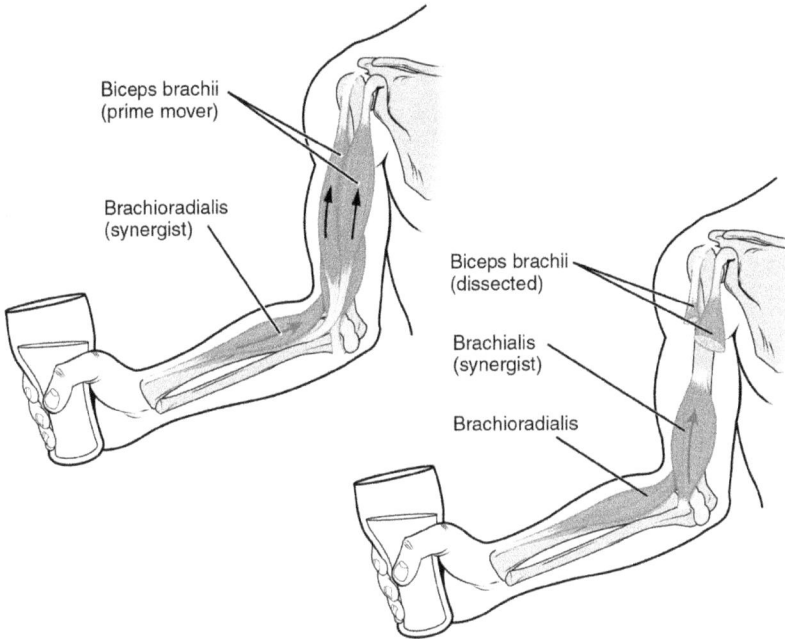

Figure 3 Textbook diagram of how contracting muscles move a bone providing an etiological mechanistic explanation. (From Betts et al. 2013, ch. 11, sec. 11.1, fig.11.2)

I will call this an Instantial explanation because it is backed by the instantiation of a property by an individual.[6]

Lastly, consider how we answer the question "Why is that individual so strong?", since one good answer, in the appropriate context, is "Because it is a skeletal muscle." Here we have a model of an individual falling under a certain kind – where the kind is being a skeletal muscle – and this kind has, among other characteristic properties (in the conditions), the property of having a certain strength. We thus explain a property of an individual using the kind of this individual and I will term this a Kind-Backed explanation.

I am terming all of the models in this subsection "intralevel" models/explanations, since these models all solely represent entities at what scientists in this area term the same "level" as each other, rather than components of these entities.[7] Having a sense of the variety of intralevel ontic models/explanations in our example, let us now turn to some of the interlevel models/explanations that researchers also offer.

---

[6] Note that the New Mechanists have plausibly not recognized either this kind of model/explanation or the next kind that I outline.

[7] By "level" here I mean what scientists in physiology, cell biology, and molecular biology mean by this term. See Gillett (2021) for a detailed account of this scientific use of "compositional level."

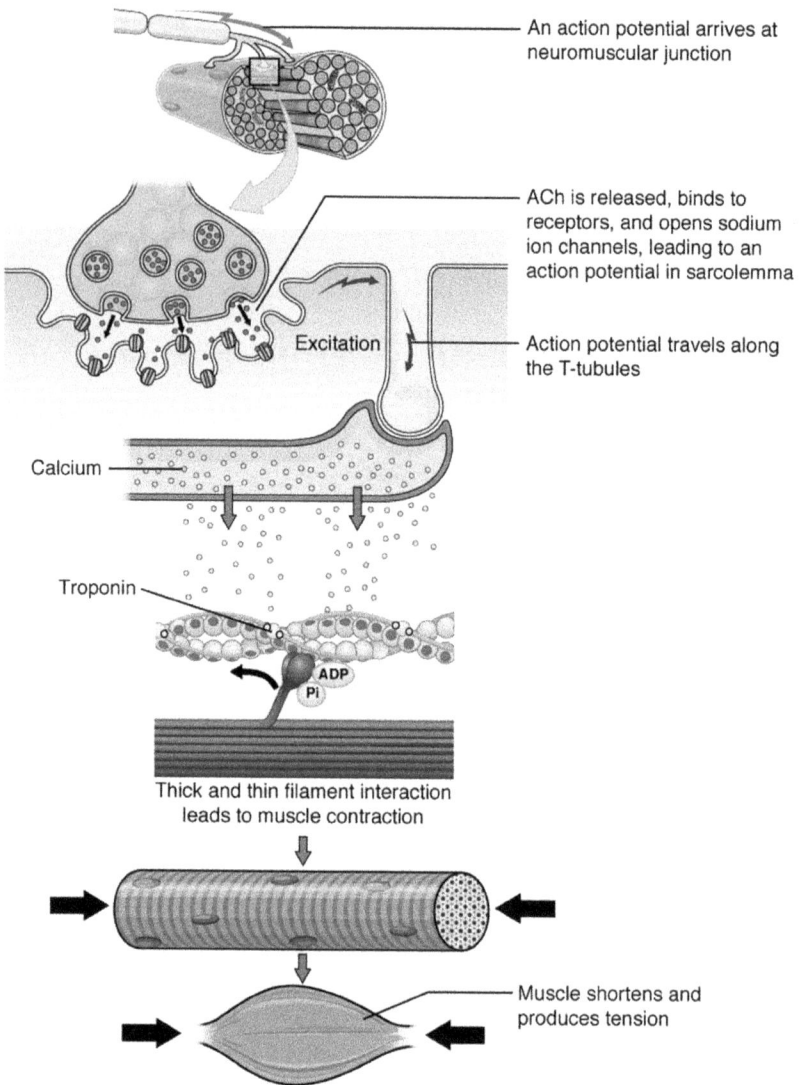

An action potential arrives at neuromuscular junction

ACh is released, binds to receptors, and opens sodium ion channels, leading to an action potential in sarcolemma

Excitation

Action potential travels along the T-tubules

Calcium

Troponin

ADP
Pi

Thick and thin filament interaction leads to muscle contraction

Muscle shortens and produces tension

Figure 4 A textbook diagram of the sliding filament model of muscle contraction and a Dynamic compositional model. (From Betts et al. 2013, ch. 10, sec. 10.3, fig. 1)

1.1.2 Some Interlevel Models/Explanations

In response to the question "Why is the muscle now contracting?" two good answers, in certain contexts, are based around the multilevel model in Figure 4 and are "Because the cell fibers are now contracting" or "Because the myosin is now crawling along the strands of actin." I term this a Dynamic compositional explanation.[8] Here we explain the muscle's contraction *at some time and place* using a compositional relation to the contraction of various cells *at the same time and place*.[9] The cells are interconnected, or "organized," so that as each contracts it pulls on the cells to which it is connected and which are also contracting. Hence the contracting cells at a certain time and place compose (or what I term implement), and explain, the muscle's contracting at that same time and place. That is, activities of parts at a certain time and place implement, and explain, an activity of a whole at that time and place.[10]

In answer to the closely related question "Why did the skeletal muscle contract over that duration of time?" two good answers are "Because the constituent muscle cells/fibers were contracting together over that duration," at the cellular level, and "Because the constituent myosin proteins were walking down the strands of actin over that duration," at the molecular level. I follow the New Mechanists in terming these constitutive mechanistic models/ explanations. In these models, we are being asked to explain a *temporally extended activity*, rather than an activity at a time which was the focus of the first question. We can reinterpret Figure 4 to represent the distinct models underlying each of the two examples of constitutive mechanistic explanations we noted. But this distinct kind of model is backed by *both* a series of compositional relations holding, at particular times and places, between activities of parts and wholes, *and* also the nature of a temporally extended series of activities of parts.

Let us move on to another kind of interlevel model also offered in the groups framed about skeletal muscles and their contraction. To the question "Why does the skeletal muscle now have energy X?" two good answers offered by researchers are "Because the combined energies of the constituent muscle cells/fibers is now X," at the cellular level, or "Because the combined energies of the constituent proteins is now X," at the molecular level. Here each of the latter explanations that I term Standing compositional models/explanations uses a model positing a compositional relation, that I will term a realization relation, between a property of the whole at a certain time and place, and properties of parts at the same time and place.

Consider another of these Standing explanations offered in the same coalition of models. To the question "Why does the skeletal muscle have strength Y?"

---

[8] Aizawa and Gillett (2019).

[9] I italicize "at the same time and place" to emphasize how the explanans and explanandum entities contrast with those of constitutive mechanistic explanations.

[10] I thus now take the Dynamic compositional, and the constitutive mechanistic models/explanations I discuss next, to be distinct because they differ in their explanans, explanandum, and backing relations in ways highlighted in Table 2. This change corrects the discussion of Aizawa and Gillett (2019).

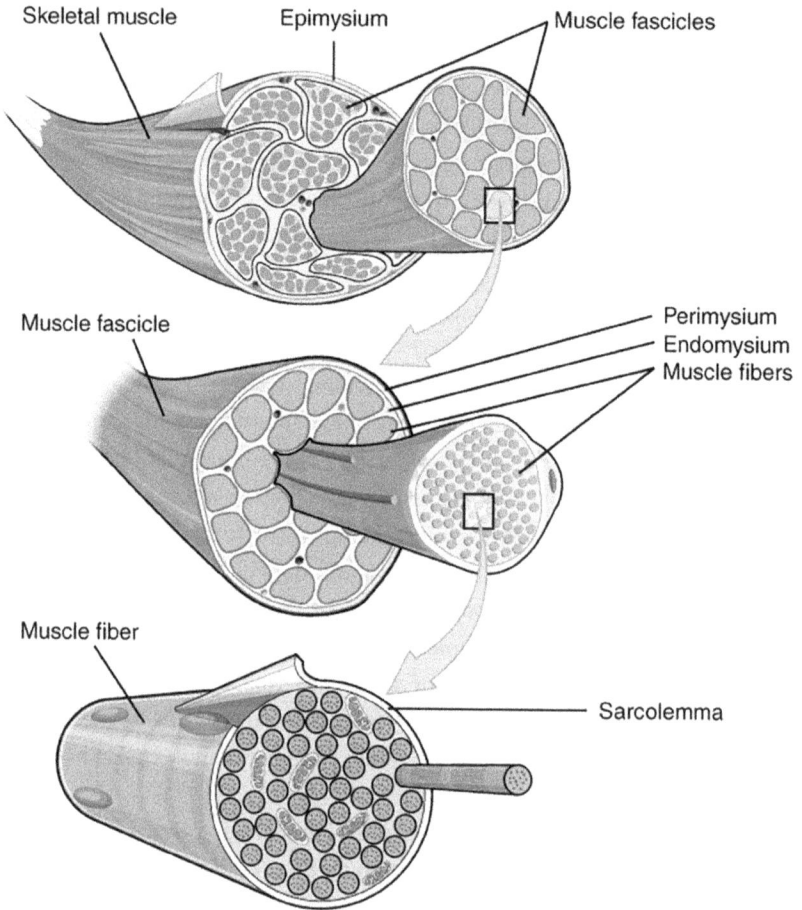

Figure 5 A textbook diagram of the composition of a skeletal muscle at tissue and cellular levels, and hence an Analytic model of it. (From Betts et al. 2013, ch. 10, sec. 10.2, fig. 1)

two good answers offered by researchers are "Because the constituent muscle cells now have certain relations and minute strengths," at the cellular level, or "Because the constituent myosin now has the property of exerting a certain force in moving down actin," at the molecular level. Again, each of the latter explanations uses a model positing a realization relation between properties of parts at a certain time and place and a property of a whole at the same time and place.

Lastly, we should note that when asked "What is a skeletal muscle?" two good answers (among others) in the relevant contexts, are, as Figure 5 highlights, "Bundled muscle fascicles," at the tissues level, or "Bundled muscle fibers," at the cellular level. Here the explanandum is a whole at a specific time and place, that is, an individual, while the explanans is some group of parts (at a certain "level") at the same time and place. And in the models underlying these explanations the backing relation is the part–whole (or what I also term a constitution) relation between these individuals. I call this species an Analytic compositional model/explanation where we explain a whole itself using a compositional relation to individuals that are its parts at a certain level.

I am terming the models/explanations in this subsection "interlevel" because each of these models/explanations are backed either solely, or in part, by compositional relations between composed and component entities at what scientists themselves take to be different levels of the body – whether part–whole/constitution relations between individuals, or realization between properties of parts and whole, or implementation between the activities of parts and whole.

1.1.3 Plural Arrays of Ontic Models/Explanations

Many philosophers of science still take scientific explanations to be exhausted by causal or mechanistic explanations, but my survey shows scientists offer other kinds of ontic model/explanation as well. For Instantial and Kind-Backed explanations are not causal. Neither the instantiation of a property by an individual, nor an individual being of a certain kind, is itself a causal relation, since they have different features from causal relations.[11] Similarly, Dynamic, Standing, or Analytic compositional models/explanations are also plausibly not causal explanations. For the backing relations of these models/explanations again require a variety of features lacking in causal relations.[12]

Instantial, Kind-Backed, and the various species of compositional explanation are also not mechanistic explanations either, since mechanistic explanations plausibly require temporally extended activities as explananda and explanantia.[13] In Instantial explanations, the explanans is simply the instantiation of a property, whilst the explanans in Kind-Backed explanations is an individual being of a kind, rather than an activity or activities of these individuals. Similarly, Standing and Analytic species of compositional explanations

---

[11] For instance, instantiation, and falling under a kind, are relations each requiring these features, among others: (a) being a synchronous relation (whose relata are also spatially overlapping) and (b) having relata involved in synchronous changes. In contrast, causal relations require neither (a) or (b).

[12] Compositional relations have distinctive features including both features (a) and (b) outlined in note 11 and also these further features not required by causal relations: (c) having relata that are in some sense the same (though not identical) and (d) being what I term a natural internal relation such that if the entities on one side of this relation are found at a certain time and place, under certain conditions, then one has the entity on the other side of the relation at that time and place.

[13] For instance, see the treatments of "mechanistic explanations" in Machamer et al. (2000), Illari and Williamson (2012), or Glennan (2017) which all overlap in this commitment.

have no kind of activity at all as explanans or explanandum, whilst the Dynamic variety has activities at some time as explanans and explanandum, rather than temporally extended activities.

Table 1 consequently outlines the distinct varieties of intralevel models/explanations, and their features, highlighted in the example examined here. And Table 2 outlines the interlevel models/explanations I surveyed here and their features.

We have thus found that researchers offer many more kinds of models/ explanations than just causal or mechanistic ones in this mature case in the context of understanding. It also bears emphasis that there are further kinds of models/explanations in this case and others like it. And other sciences, focused on different questions, use still other kinds of model/explanation using internal ontologies with very different ontological categories.

Our models/explanations from physiology all focus on a historically individuated entities, including what New Mechanists term working individuals individuated by their activities, what we may term working properties individuated by their powers to engage in such activities and various activities themselves. In contrast, for example, evolutionary biology often posits historically individuated entities in its models/explanations which are not working entities of this kind.

However, for my purposes here it suffices to highlight how scientists in physiology, cell biology, and molecular biology offer such a plural array of ontic models. Interestingly, these models are also plausibly systematically integrated with each other in ways that allow them to supplement each other. I explore this interesting characteristic further in a subsequent subsection. But, in the next subsection, let me first consider the neglected family of compositional models/explanations our survey highlights.

1.2 Compositional Models/Explanations in the Sciences: A Neglected Family

Our survey begins to showcase the family of compositional models/explanations in the sciences. A few philosophers of science and mind have focused on these models/explanations, but mainstream philosophy of science has largely ignored or denied these models/explanations.[14] Let me therefore give a brief

---

[14] See, for example, Fodor (1968) and Dennett (1978), who pressed cognitive science to use such compositional models/explanations. Wimsatt (1976) and (2007) has long acknowledged such models/explanations and the levels associated with them. And, more recently, work such as Aizawa and Gillett (2019), Gillett (2007a, 2016a, 2021, 2022, and Unpublished), Love (2012), and Love and Huttemann (2011), amongst others, have all begun to focus on such compositional models/explanations.

Table 1 Summary of some of the intralevel models/explanations we find in practice and their differing characteristics.

Species of Model/ Explanation	Represented Categories of Entity	Explanandum	Explanans	Backing Relation
(Simple) Etiological Mechanistic Model/Explanation	Individuals and activities	An activity or property of some individual	An activity of an individual	A thick causal relation of activity
Instantial Model/Explanation	Individuals, properties, and activities	An activity of an individual	A property of the same individual	An instantiation relation between the individual and the property
Kind-Backed Model/Explanation	Individuals, properties, and kinds	A property of an individual	The kind of the same individual	The individual falling under a kind

Table 2 Summary of some of the interlevel models/explanations we find in practice and their differing characteristics.

Species of Model/ Explanation	Represented Categories of Entity	Explanandum	Explanans	Backing Relation
Constitutive Mechanistic Model/Explanation	Individuals and activities	An activity of a whole over time	Activities of individuals that are parts over time	Implementation relations between activities of parts and whole at certain times, and relations between the activities of parts over time
Dynamic Compositional Model/Explanation	Individuals and activities	An activity of a whole at a time and a place	Activities of individuals that are parts at a time and place	Implementation between activities of parts and whole
Standing Compositional Model/Explanation	Individuals and properties	A property of a whole at a time and place	Properties of parts at a time and place	Realization between property instances of parts and whole
Analytic Compositional Model/Explanation	Individuals	An individual that is a whole at a time and place	Individuals that are parts at a time and place	Constitution or part–whole relation between individuals

sketch of these models/explanations, and their backing relations, since they are a central player in the scientific debates we will examine in coming sections.

This family of models/explanations are backed by compositional relations that I have noted are very different from the causal relations that have dominated discussion for so long in contemporary philosophy. Causal relations, which back scientific models/explanations, usually hold over time between relata at different times and places where these relata are wholly distinct entities and where there is a transfer of energy and/or mediation of force. In contrast, compositional relations hold between relata at the same time and place. Furthermore, the relata of compositional relations are in some sense the same (but not identical) and thus do not transfer energy or exert force upon each other. Crucially, and unlike causal relations, compositional relations are also what I term natural internal relations because when one has the component entities at a certain time and place, under specific conditions, then one also has the composed entity at that time and place.

Putting things more simply, a causal relation holds between distinct things at different places where the relation between these entities unfolds over time using energy and/or force. Thus, a skeletal muscle contracts over time to pull on a bone and change the position of this bone during this period of time. In contrast, a compositional relation involves components at some time and place, under certain conditions, that synchronously result in the composed entity at that same time and place without a transfer of energy, and/or exertion of force, to mediate this relation.

To use an older phrase, the components thus provide a reason for existence of the composed entity and hence explain it. For example, in one of our examples of a Dynamic compositional explanation we saw how when there are many myosin proteins crawling along actin filaments, at a certain time and place under specific (and very complicated) conditions, then one also has a contracting skeletal muscle at the same time and place. Under the conditions, we can thus explain the muscle's contracting at a certain time and place using the compositional relation to these activities of its constituent proteins at that same time and place.

We should also carefully mark that in our survey we found that scientists use various *species* of compositional model backed by *distinct* compositional relations involving *different* categories of entity (see the last column of Table 2 for a summary.) Furthermore, we saw that the species of compositional model, and their distinct compositional relations, are offered together. For instance, the individuals taken as parts in our Analytic compositional models/explanations are just the individuals whose activities, and properties, were taken to implement activities, and realize properties, of the same kind of whole in Dynamic and Standing compositional models/explanations. The distinct compositional backing relations of the three kinds of models are thus systematically and tightly interconnected. In the next subsection, I highlight reasons why scientists might seek such systematically integrated models, and backing relations, but to finish our brief discussion we need to discuss the general nature of such scientific composition relations.

More detailed theoretical accounts need to be provided of such scientific relations. And there is now a flowering of broader philosophical work on a variety of "vertical" relations that one might seek to use to do this.[15] However, extreme care needs to be taken before one simply appropriates philosophical accounts of "vertical" relations and blithely applies them to the compositional relations in scientific models/explanations. For these philosophical notions were not crafted to apply to scientific composition relations which plausibly differ in their features. Elsewhere I have made these negative points in detail, but I leave them to the side here.[16]

Instead, to use in coming sections, let me simply offer a positive thumbnail sketch of the compositional relations in the models/explanations we surveyed. To start, as I noted earlier, mark that all the entities in the models/explanations in our cases are ahistorical, working entities, to use the New Mechanist's term, since these entities are individuated, in one way or another, by the actual or potential activities they are connected to. We can thus crudely use "roles" framing their associated activities to individuate working entities, whether individuals, activities, or properties.

Against this background, I suggest components are entities, at a certain time and place, and under specific conditions, whose roles together result in the role that is individuative of the composed entity – and hence result in the composed entity at that time and place. I will call this kind of relation "joint role-filling." For example, a team of interrelated molecules, including actin, myosin, and so on, jointly fills the role of a muscle cell. Or a team of various activities of muscle cells, and other cellular activities, jointly fill the role of the activity of contracting in a muscle cell. Or the energies of various cells jointly fill the role of a muscle's property of having a certain energy. Under this joint role-filling account, scientific components are thus members of teams of many working entities spatially contained within one composed entity and interrelated such that, under the conditions, they jointly fill the role of – and hence result in – the composed entity at the same time and place. There are now detailed joint role-filling accounts of

---

[15] For a survey of the three broad traditions of philosophical work on "vertical" relations, across the areas of philosophy of mind, philosophy of science, and analytic metaphysics, see Aizawa and Gillett (2016b).

[16] Gillett (2016a, ch. 2), and Gillett (2016b).

each of the three distinct, compositional relations we find in scientific models/explanations.[17] But, again, I leave those detailed positive accounts to the side here. For my present purposes, the rough general characterization of composition as a joint role-filling relation will suffice.

1.3 Integrative Pluralism and the Synthetic View of Explanation

It is striking in our case that researchers offer many distinct, but integrated, models/explanations to better understand the same state of affairs. These groups of plural, but integrated, models in the context of understanding are what I am terming "coalitions." We thus have an example of what Sandra Mitchell (Mitchell 2003) has termed integrative pluralism. One might well wonder *why* researchers offer such plural but integrated groups of models/explanations about the same natural phenomenon? Mitchell has offered the basis of a compelling answer that we can add to.

A lesson taught to us by recent debates about models in the philosophy of science is, as Mitchell notes, that researchers make complex phenomena cognitively tractable by only *selectively representing*, and addressing, certain aspects of the phenomena in any one model. Tables 1 and 2 detail how each of the kinds of model in our case only selectively represents a few, differing, categories of entity. The strategy behind using such selective representations is apparently "Divide and Cognize," rather than "Divide and Conquer," to make complex states of affairs cognitively tractable.

Using this strategy, researchers thus end up with a plural array of models, each of which selectively represents differing categories of entity that allows each model to successfully explain a distinct facet of the same state of affairs. Researchers then integrate these models/explanations, since these integrated, selective ontological representations *together* provide a more comprehensive understanding of the state of affairs than any single selective model/explanation.

Crucially, in our example we can therefore see that we have plural *selective*, but also *supplementive*, representations. When these plural models are integrated, then together these models supplement each other to offer a more comprehensive understanding of some complex state of affairs in nature. Using coalitions of models thus has a solid rationale.

Let us now briefly consider the manner of integration of such coalitions. What I suggest we find is that these models are integrated through systematically

---

[17] For example, joint role-filling accounts of the "realization" of properties of parts and whole are offered in Aizawa (2007), Gillett (2002, 2016a, Unpublished), Pereboom (2002, 2011), and Shoemaker (2007); for the "constitution" and "part–whole" relations of individuals see Gillett (2007b, 2013, 2016a, Unpublished), Kaiser (2018), Glennan (2020), and Pereboom (2011); and for the implementation of activities of parts and whole see Gillett (Unpublished, ch. 5).

interconnecting their "internal ontologies" – that is, the entities in nature that each model represents. (I use the term "ultimate ontology" to mean entities in nature itself.) Such integration takes many forms. For instance, we should notice that the models in our example all systematically overlap, and mesh, in the kinds of entities that they represent at various levels. Furthermore, we have still more complex interconnections between the internal ontologies of these models. For example, as I noted earlier, the three kinds of compositional relations backing the species of compositional model/explanation in our case are all tightly inter-twined along with their relata. And there are still other interconnections of this type.[18]

Rather than a chaotic jumble of models positing different kinds of individuals, engaging in unrelated or even incompatible activities, or having unrelated or incompatible properties, the distinct kinds of selective representations offered by scientists in our case posit kinds of entity that are systematically interconnected to the entities represented in the other models in a coalition. The result is a group of models that each supplements the understanding that the other models provide so that the coalition together gives a more comprehensive understanding of the state of affairs in nature – hence providing a distinctive manifestation of Mitchell's integrative pluralism in the context of understanding.

The importance of such interconnections between the internal ontology of different models is magnified if we briefly note the character of the models/explanations we have surveyed. I suggest that what I term the "Synthetic" view best captures these models/explanations. Under the Synthetic view, explanations are representations, in the form of models, of various entities in nature which have explanatory power, that is, "explanatoriness," about a certain natural phenomenon as a result of the character of the entities in nature that they represent. I term this the Synthetic view because it synthesizes key insights of the two currently dominant views of explanation. It accepts the insight of the so-called Epistemic view, defended by writers like William Bechtel (2005) and Cory Wright (2012), that such explanations are representations. But it also embraces an insight of the opposing Ontic views, pressed by writers like Wesley Salmon (1989) and Carl Craver (2007), that the character of represented entities in the world drive an explanation and underlie its explanatory power.

Under the Synthetic view, models (and hence explanations) are representations that get their content – what they are about – by being interpreted in certain ways by the relevant scientists (Giere 2010) using their ontological concepts (whether of individuals, activities, properties, compositional relations, kinds,

---

[18] See Gillett (2021, 2022, and Unpublished) for more detailed discussion of these fine-grained interconnections among the internal ontology of such models.

etc.). Consequently, the *ontological concepts* of researchers inform, and in fact underwrite, their models and hence the explanations that these models provide. We can thus begin to see why endogenous metaphysics is central to foundational scientific practices, since the ontological concepts of scientists provide content, and explanatory power, to their models.

1.4 The Dynamic Cycle of Ontic Concept/Model Creation, Application, Assessment and Alteration

The ontological concepts of scientists are, perhaps unsurprisingly, central to their models/explanations in the context of understanding. And, once we appreciate integrative pluralism holds true of these models, as Mitchell (2003) notes, we can expect a changing story about the kinds of models/explanations researchers use at different periods as they develop new types. That prompts the general question of *how* researchers develop new ontological concepts and the novel models/explanations they underwrite?

The latter is a large and important question, but we need some answer in order to move our examination of reduction–emergence debates forward. I therefore take a rough stab at sketching an abstract answer in what I am terming the Dynamic Cycle depicted in Figure 1. As we have seen, the models/explanations we have looked at in our case are representations underpinned by ontological concepts which result in the model's "internal ontology." And someone needs to develop those ontological concepts, and hence models – whether of a certain kind of individual, or the specific activity it engages in, or the property this results from – where this is a substantive, and creative, intellectual task.[19]

This latter kind of work, in the top right stage of the Dynamic Cycle, goes on in the context of understanding where the primary goal of researchers is to develop concepts to construct models to describe, explain, and understand certain states of affairs in nature.[20] But the Dynamic Cycle does not end with such theoretical work focused on representing and understanding.

In the next stage of the Dynamic Cycle, at the bottom right of Figure 1, such models/explanations are then applied in the context of investigation where the goal of scientists is to empirically explore states of affairs in nature to produce observations, measurements, experimental results, and other empirical findings about these states of affairs. Ontic models/explanations are used to, and are

---

[19] Furthermore, researchers also ultimately need to integrate the internal ontologies of their various models, too. See Sullivan (2017)'s discussion of the "coordinated pluralism" required for that still more complicated task that is plausibly also driven by the Dynamic Cycle.

[20] Recent work on a range of actual cases, I contend, supports the use of the Dynamic Cycle. See, for example, Bollhagen (2021), which, charitably interpreted, nicely illustrates the Dynamic Cycle in practice in physiology, cell biology, and molecular biology.

plausibly necessary for, reliably designing successful experiments or other investigative tools that are used in the context of investigation. This stage in the cycle thus often involves applications of models/explanations to work focused on "intervening" in nature. So, for instance, researchers might design experiments or tools to produce findings about the activities, or properties, represented in their models.

The Dynamic Cycle thus spans *both* the context of understanding, where ontological concepts, models, and explanations are created with the primary goal of understanding natural phenomena, *and also* the context of investigation, where such ontic models/explanations and others are applied to nature through the design of experiments and other investigative tools to generate empirical findings which are often relevant to these models.

The latter points are significant because once researchers have relevant empirical findings, then they can eventually move to the third stage, in the lower left of Figure 1, which returns to the context of understanding. Using these findings and other relevant evidence, researchers assess how well (or badly) their various models, and their underlying ontological conceptions, succeed in accurately describing, explaining, or otherwise allowing us to understand the relevant states of affairs in nature. That is, researchers assess how well the internal ontology of their models/explanations reflects what we know of the ultimate ontology of nature itself. This stage of the Dynamic Cycle thus plausibly involves researchers making decisions about both the internal ontology of their models *and* our evidence about the ultimate ontology of nature.

That takes us to the last stage in the Dynamic Cycle, in the upper left of Figure 1, where scientists decide how to react to their assessment of their extant ontic models and the ontological concepts that underwrite them. A scientist might choose to revise their initial ontological concept to one that better fits with the empirical findings. Alternatively, scientists can supplement their extant models with new ontological notions underwriting still further models to supplement them. Or researchers can choose to completely replace existing models, and their ontological notions, with replacements underwritten by a different internal ontology. These decisions will obviously vary greatly depending upon the case and the evidence.

Each of those decisions, about revising, supplementing, or even replacing their ontic concepts and models/explanations, then leads researchers back to the stage of creating new ontological concepts and models in the upper right-hand stage of Figure 1 – and to a new turn of the Dynamic Cycle. Whatever researchers decide, the resulting models/explanations constructed in the context of understanding will then once more be applied in the context of investigation and assessed in light of its empirical findings.

The Dynamic Cycle is plausibly put to work at the local level where researchers revise, and shape, existing concepts – and hence models – of individuals, or their properties or activities, in light of empirical findings. Such local work is often incremental and involves altering existing ontological concepts to produce more adequate models/explanations. Illuminating how the Dynamic Cycle works locally in this manner is an exciting area of ongoing research for philosophers of science.

However, for my purposes here, what is more important is that appreciating the Dynamic Cycle opens a still further possibility – namely, that scientists might sometimes pursue *global* and/or *categorial* ontological innovations to underwrite whole new families of model. Such categorial innovation would appear to be most likely when local, incremental innovations have consistently failed to address long-standing, and seemingly intractable, scientific problems – hence prompting a more radical approach. For scientists experimenting with new ontological categories can potentially underwrite novel families of ontic models/explanations that may provide fresh resources with such tough problems.

If a group of researchers were indeed to pursue such global ontological innovation and pioneer a whole new family of ontic models, then the usual collaborative and coordinative aspects of the Dynamic Cycle would presumably require a greatly magnified form of what Sullivan (2017) terms "coordinated pluralism." To that end, researchers across various research groups, areas, and even sciences, would have to be educated about the relevant ontological innovation and its implications. To communicate what their view says about the structure of nature, I assume such groupings will outline what I term the "guiding picture of nature" provided by their view in a broad, easily communicable account that allows other researchers to grasp the nature, and scientific import, of the relevant ontological innovation. For researchers will also have to be educated about both the novel models and methodologies resulting from the position.

Researchers involved in such efforts, in order to deal with the required coordination, can thus be expected to often organize themselves into a broader social grouping that I will term a "Global Ontological Research Movement" – that is, like-minded scientists seeking to pioneer and disseminate a global ontological innovation and the resulting new models and methodologies. Such research movements obviously draw inspiration from Larry Laudan's (1977) work, but I give them a different name because they may occur within the still bigger scientific "units" that Laudan himself terms "research traditions."[21]

---

[21] Note that there could be other kinds of research movements grouped around promoting the use of new investigative tools, or mathematical or representational techniques, amongst other possibilities. (See Bickle et al. 2022 for discussion of possible tool-based movements.)

In coming sections, I explore the possibility that the two cycles of reduction–emergence debates in the sciences involve researchers in such Global Ontological Research Movements. For such debates have featured broader groupings of researchers, self-identifying as "reductionists" or "emergentists," who each press novel pictures of the ontological structure of nature that underwrite new models/explanations, and/or novel methodologies, in response to long-standing scientific problems. Hence apparently pursuing the Dynamic Cycle. I also use our examination of the positions in the two cycles of reduction–emergence debates to sketch the nature and features of the movements that we find in actual scientific practice.

1.5 Exogenous and Endogenous Metaphysics

Before we explore reduction–emergence debates, however, we need to finally confront an assumption of philosophers of science that has plausibly hindered them in this endeavor. For many mainstream philosophers of science assume there is just one kind of metaphysics that they take to be pursued by contemporary analytic philosophers in the area labeled "metaphysics."[22] I should carefully mark that I am *not* endorsing the accuracy of this picture, and, in Section 5, I suggest there is plausibly a variety of approaches in analytic metaphysics. Instead, I am simply framing the assumptions of philosophers of science to bring out why they may have overlooked a different kind of metaphysics and the scientific practices involving it.

Philosophers of science take their sole exemplar of metaphysics, that they see in analytic philosophy, to use *a priori* methods to explore folk concepts, or concepts from formal frameworks, to draw conclusions about ultimate ontology with no engagement with the products of the sciences, their internal ontology, or associated empirical findings. The intended users of the resulting accounts in this kind of metaphysics, as Figure 6 depicts, are also taken to be other researchers in theoretical debates within philosophy. Accounts offered in this kind of metaphysics thus dead-end in such theoretical debates and are hence never applied to, nor empirically tested against, reality. Consequently, issues pursued in this type of metaphysics are taken to be unresolvable empirically. I term the latter "exogenous" metaphysics because it is taken to be practiced outside the sciences with little contact with their products or findings.

Assuming that metaphysics must be of this exogenous variety is extremely unfortunate because the work of this section begins to highlight how endogenous metaphysics, pursued in the sciences, is starkly different.

---

[22] For a vivid example of this understanding of "metaphysics," see the early chapters of Ladyman and Ross (2007).

A priori investigation of folk or formal concepts	⟹	Construction of new ontological concepts, arguments, explanations and/or positions	⟹	Assessment of accounts (and possibly revision) solely within theorectical debates in philosophy

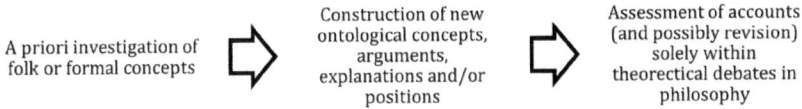

Figure 6 Abstract diagram of the received view, in philosophy of science, of metaphysics and its stages in analytic philosophy as a process dead-ending in theoretical debates in philosophy with no engagement with scientific investigations and their findings or other empirical evidence.

Endogenous metaphysics, as we saw in earlier subsections, uses *a posteriori* methods and focuses on ontological concepts intended to underwrite ontic models/explanations built to be used within science. The users are working scientists who construct models/explanations in the context of understanding and then apply these models/explanations in the context of investigation. Rather than dead-ending in theoretical debates in the context of understanding, such researchers following the Dynamic Cycle thus apply their ontological concepts/models in the context of investigation in empirical work, assesses these models using the resulting empirical findings, and then revise, supplement, or replace these concepts/models in further iterative turns of the Cycle. Consequently, issues in endogenous metaphysics are often empirically resolvable – as we shall see in the next section.[23]

Contrary to the assumptions of many philosophers, exogenous metaphysics is therefore plausibly not the only variety of metaphysics. Endogenous metaphysics, underlying the Dynamic Cycle and other foundational scientific practices, is another kind pursued for a distinct purpose, involving the different practices and characteristics, framed in Table 3.

Many philosophers of science have assumed exogenous metaphysics is the only variety and, given its features, have hence endorsed a blanket dismissal of all "metaphysics" as relevant or fruitful for scientific research. But this has often left mainstream philosophy of science blind to the endogenous metaphysics central to many foundational scientific practices. We therefore plausibly need to look again at a variety of scientific phenomena involving endogenous metaphysics to reassess whether mainstream accounts in philosophy of science have accurately reflected the scientific realities – or not. In subsequent sections, I begin the project of reassessment for the two cycles of reduction–emergence debates in the sciences.

---

[23] For what appear to be related pictures of the metaphysics done within science see, for example, Dupre (2021) and the papers in Andersen and Mitchell (2023).

Table 3 Summary of the characteristics of the common picture of analytic metaphysics and a distinct variety of metaphysics in the sciences.

Type of Metaphysics	Focus	Methods	Intended End-User	Shape of the Practice	Empirically Resolvable Issues
The Exogenous Metaphysics in Analytic Philosophy (as conceived by philosophers of science)	Ultimate ontology	*A priori* reflection on folk or formal concepts with little engagement with the sciences or their products, their empirical findings, and their internal ontology	Analytic metaphysicians in theoretical debates in philosophy	Dead-ends in theoretical debates in philosophy with no systematic contact with, or assessment using, empirical evidence (depicted in Figure 6)	Issues are not usually empirically resolvable
The Endogenous Metaphysics in Science	Internal and ultimate ontology	*A posteriori* investigation of models or explanations, and their underlying ontological concepts, in the sciences	Scientific modelers engaged in scientific inquiry about states of affairs in nature	Innovative, iterative, and potentially progressive cycle of concept/model construction, assessment, and revision in scientific practices involving empirical application and evidence (depicted in Figure 1)	Issues are usually empirically resolvable

1.6 Global Ontological Research Movements in the Sciences?

One critical response to the endogenous metaphysics I have highlighted in this section is that objectors claim it is no metaphysics at all. For, these objectors contend, such scientific practices are simply the local activities of researchers revising existing ontological conceptions of entities, or at most creating new conceptions of extant categories of entity (whether individuals, activities, properties, kinds, etc.), rather than the exploration and development of new ontological *categories* that is the proper focus of any metaphysics.

I obviously agree with one presupposition of this objection – namely, that developing new ontological concepts of entities of various categories is central to foundational practices in ordinary science. But I do not want to get bogged down in often fruitless debates over what metaphysics "really" is. Instead, in coming sections, I want to press ahead and look again at the two sets of reduction–emergence debates in the sciences using the alternative platform provided in this section. For we have now seen that there is a real danger that philosophers of science have wrongly dismissed metaphysical debates, as having no relevance for scientific practice, based on an overly narrow view of what metaphysics must be.

Once we re-examine the reduction–emergence debates, I show that such dangers are real. For I outline how in each cycle of reduction–emergence debates we have plausible examples of Global Ontological Research Movements pioneering *categorial* ontological innovations to underwrite new families of models/explanations to speak to the problems in that period. Crucially, I also show that extant kinds of scientific reductionism and emergentism pass our objector's own bar for metaphysics, since these scientific movements do in fact involve exciting innovations in ontological *categories* – albeit to underwrite new families of model to solve scientific problems.

2 The Twentieth Century's Great Reductionist Movement and Its Search for Compositional Explanations

Reductionism (Oxford English Dictionary) Definition: The practice of describing or explaining a complex (esp. mental, social, or biological) phenomenon in terms of relatively simple or fundamental concepts, especially when this is said to provide a sufficient description or explanation; belief in or advocacy of such an approach.[24]

Our starting dictionary definition flags that in the wider intellectual world, beyond philosophy, "reductionism" has often been used to mean a scientific practice of describing or explaining wholes in terms of their parts. And this is

---

[24] *Oxford English Dictionary* (2022).

no accident. For the great reductionist movement of twentieth-century science – what I am terming Everyday Reductionism – focused on searching for and constructing compositional models/explanations across a range of disciplines and levels of nature. Sadly, philosophers of science have largely overlooked this movement given their continuing blindness to compositional models/explanations.[25] I use this section to start to rectify our situation by providing a sketch of Everyday Reductionism and its role in twentieth-century science.

I begin, in Section 2.1, by outlining the first cycle of reduction/emergence debates in the sciences of the late nineteenth and early twentieth century which focused on long-standing problem cases in chemistry and biology. I highlight how on one side were the dominant "emergentist" views of this period that argued that the relevant chemical, biological, and other higher-level phenomena were uncomposed and hence involved fundamental, or so-called special, forces and/or energies. On the other side of this debate, I note that the most prominent reductionists were Everyday Reductionists who pressed a novel, integrated view under which all working phenomena in nature, above physics, were composed.

Given its historical importance, I focus on the Everyday Reductionist side of this first debate. In Section 2.2, I sketch the categorial ontological innovation of Everyday Reductionism which takes not only all working individuals to be composed, but also all of their activities and working properties as well. Perhaps more importantly, I sketch how this ontological innovation underwrites the new family of integrated compositional models/explanations, highlighted in the previous section, that compositionally explains individuals and their activities and properties, too. I outline how Everyday Reductionism consequently supplies researchers with new resources both generally as well as in the chemical or biological problem cases.

Given these features, I conclude, in Section 2.3, that Everyday Reductionism is the type of Global Ontological Research Movement that I predicted might be found in the sciences. I also suggest Everyday Reductionism highlights the characteristic features of such a movement in (i) a global ontological innovation which underpins a new guiding picture of nature; (ii) a new family of ontic models/explanations, underwritten by (i), that are applicable across many sciences; and (iii) new methodologies driven by (i) and (ii).

---

[25] The Everyday Reductionist movement, as this section lays out, is neither the philosopher's Nagelian reductionism seeking to derive laws of higher sciences from laws of lower science, nor is it a version of the philosophically familiar ontologically reductionist positions (what I term Fundamentalism in the next section) arguing that there are parts alone.

Having confirmed that we do indeed find Global Ontological Research Movements in scientific practice, another question immediately arises: Have such movements ever been *successful*?

I suggest, in Section 2.4, that Everyday Reductionism again allows us to answer this question with a resounding "Yes." I sketch how Everyday Reductionism was one of the most successful research movements of twentieth-century science – and for a long time what scientists meant by "reductionism." For scientists piled up an array of headline-making and mundane compositional models/explanations through the twentieth century across many sciences. Given this success, by the middle to late twentieth century, the first round of reduction–emergence debates were taken to be resolved in the sciences in favor of Everyday Reductionism. And what philosophers term "physicalism" about nature was endorsed by scientists – hence highlighting how issues in endogenous metaphysics are indeed empirically resolvable. I therefore suggest, in Sections 2.4 and 2.5, that all scientists are now consequently Everyday Reductionists and that the findings of this movement provide a crucial background to our present debates.

2.1 The First Set of Debates and their Battles over Compositional Explanation

Philosophers are familiar with the early twentieth-century debates between reductionists and emergentists from the work of writers like Brian McLaughlin (1992). At the end of the nineteenth, and on into the early twentieth century, there were long-standing failures to understand higher-level phenomena using lower-level ones. For example, in chemistry there continued to be problems understanding the properties of substances through relations to the properties of their constituents. Thus, there was a long-standing and well-known failure to explain properties of common salt (NaCl) using the properties of its constituent sodium and chlorine atoms. And, in the biological sciences, there was a widely known failure to understand the activities and properties of biological individuals in terms of the activities and properties of their constituents. For instance, the digestive activity of the stomach had long resisted understanding in terms of the activities and properties of its parts.

After discussing the Dynamic Cycle, I suggested in the previous section that one way to react to such long-standing problems is to ontologically innovate in a broader categorial and/or global fashion. And plausibly two rival strategies of just this kind were pressed by the most prominent positions on either side of this first cycle of debates.

On the "emergentist" side, we find researchers who pioneered the ontologically innovative view that many higher-level individuals and their activities and properties are uncomposed and involve "special," that is uncomposed and hence fundamental, forces and/or energies. Most prominent amongst such writers, were Vitalist biologists like Hans Driesch (1929), and others, who argued that activities, like the digestion of the stomach, involved uncomposed and fundamental biological forces and/or energies. Given this ontological commitment to new, uncomposed entities, I term these Ontological emergentists (Gillett 2002a, 2016a, ch. 4.)

On the "reductionist" side, the most prominent scientific view espoused the ontologically innovative position that all working individuals, and also their activities and working properties, are fully composed by lower-level individuals and their activities and properties. This is what I am terming Everyday Reductionism, for reasons that will become clear as the section progresses, though it is often also termed in this period mechanism, or materialism, or mechanistic materialism. Crucially, using its ontological innovation, the position defends the claim that chemical, biological, and other entities are all amenable to the suite of compositional models/explanations sketched in the previous section.

I should flag that the first cycle of reduction–emergence debates had more positions defended by scientists and philosophers than the two dominant views I have just outlined. On the reductionist side of debates, for example, we also find a more radical position held by scientists like Eddington (1928) who famously implied that there are *parts alone*, rather than parts *and* wholes as Everyday Reductionism implies. Similarly, on the emergentist side of the debates, the philosopher Samuel Alexander (1920), along with organicist biologists like Joseph Needham (1936) and Joseph Woodger (1929), are all lumped together with Ontological emergentists like Broad (1925) under standard treatments such as McLaughlin's (1992). But emergentist writers like Alexander, Needham and Woodger, among others, were unlike Broad because they took everything to be composed and then further claimed that parts behave differently in wholes in ways "downwardly" determined by these wholes.[26]

Although minority players in the first cycle of debates, in coming sections I show that these other reductionist and emergentist views came to dominate in the *second* cycle of reduction–emergence debates. Why did Everyday Reductionism and Ontological emergentism dominate in the first set of reduction–emergence debates in the sciences, whereas these other views only came to the fore in later debates?

---

[26] These writers hold a position close to that held in contemporary debates by that I term Mutualism and outline in Section 5. See Gillett (2006) for a reconstruction of Alexander's position which plausibly applies to the views of Needham and Woodger. Haraway (1976) provides a detailed overview of the commitments of these organicist scientists.

Appreciating the Dynamic Cycle offers one suggestion: In the early twentieth century, our empirical evidence was such that the two dominant positions had sufficient empirical support for their central claims and hence offered models/ explanations speaking to the live issues of the sciences at this time, while the minority positions lacked these attributes. I suggest in coming sections that the resolution of the first set of debates in the sciences provided the materials to make these minority views live later in the century in our present reduction– emergence debates. In the remainder of this section, however, I focus on illuminating the character of the historically important Everyday Reductionist position in more detail, while I leave Ontological emergentism and these other views to the side.

2.2 The Commitments of Everyday Reductionism

I outline the ontological innovation of Everyday Reductionism, in Section 2.2.1, and how it goes beyond earlier materialism to give us both a new picture of nature to guide research and a new suite of models/explanations underwritten by this innovation. In Section 2.2.2, I then note how the methodologies offered by Everyday Reductionism both help with the long-standing scientific problems, but also aid in broader research projects as well. In Section 2.2.3, I conclude that Everyday Reductionism is plausibly a Global Ontological Research Movement and I highlight its important features.

2.2.1 A Novel Ontological Innovation, a New Guiding Picture of Nature, and a Suite of Integrated Models/Explanations

Advocates of materialism in the sciences had been numerous since at least Newton. However, Everyday Reductionism embraces a more detailed version of such a materialist view of nature where all working individuals, and *also all of their activities and working properties*, are composed of lower-level working individuals and their activities and properties. This picture of nature is founded upon the intertwined compositional relations underwriting the Everyday Reductionist's suite of integrated models that we surveyed in Section 1.

Dynamic compositional explanations explain any activity of a whole using compositional relations to activities of its parts. Standing explanations explain any working property of a whole using compositional relations to properties of its parts. And Analytic compositional explanations explain any working whole using a compositional relation to its parts which are those individuals whose activities and properties compose activities and properties of this whole.

Crucially, this integrated suite of models/explanations, and the systematically interconnected compositional relations that underwrite them, allows us to provide, at a time, a comprehensive compositional understanding of a working individual and *all* of its activities and working properties – hence, if successful, leaving no room for uncomposed or "special" forces or energies.

2.2.2 Promising New Methodologies: Addressing Narrow and Broad Projects

Everyday Reductionism provides an obvious methodology: Search for compositional models/explanations. But to appreciate the promise of this methodology, we need to highlight some singular features of its integrated compositional models/explanations. This illuminates why the Everyday Reductionist's methodology has promise not only with the problem cases inspiring this movement, but also more broadly in relation to working entities in many higher sciences.

Looking to the long-standing problems cases in chemistry and biology, we should note that compositional models/explanations have the characteristic I have elsewhere (Gillett 2016a, ch. 2) termed "Ontologically Unifying Power" (OUP). That is, in a situation where we initially had what we thought were independent and wholly distinct entities, we can show these entities are, in fact, in some sense the same (though not identical) by supplying successful compositional models/explanations.

Standing compositional explanations provide a nice illustration of such Ontologically Unifying Power. One might initially think that the energy of an organism, or one of its organs like a stomach, is independent of the energies of its cells or molecules. However, as work in biology in the twentieth century showed, successfully supplying Standing compositional models/explanations allows us to explain the energy of a whole, whether an organism or organ, using a realization relation to energies of its parts, whether cells or molecules. Similar points about OUP apply to all the varieties of compositional model. And Everyday Reductionism's suite of compositional models/explanations thus offered a highly promising methodology to address the long-standing problems at the beginning of the twentieth century.

Looking more broadly, we should mark that the methodology of Everyday Reductionism offers wider benefits as well. We can appreciate some of these by noting a second characteristic of compositional models/explanations in what I have elsewhere termed (Gillett 2016a, ch. 2) the "Piercing Explanatory Power" (PEP) of these models/explanations. For compositional models explain working entities of one kind using other qualitatively distinct kinds of working entity. For example, we explain the muscle's activity of contracting using compositional relations to proteins that move along, and pull on, other proteins –

hence we explain a whole's activity using the qualitatively different activities of parts. And similar points apply to the properties of wholes and parts.

Again, PEP is distinctive of all compositional models and the result is an attractive kind of explanatory power, since explaining one kind of entity using different kinds of things is a powerful tool with very broad application across many sciences.[27] Everyday Reductionism thus offers an attractive methodology with many scientific applications beyond the initial problem cases in chemistry and biology.

2.2.3 A Global Ontological Research Movement and Its Characteristics

The work of this subsection plausibly shows that Everyday Reductionism is a Global Ontological Research Movement of the kind I suggested that we might expect if scientists use the Dynamic Cycle. For we have now seen that Everyday Reductionism has these characteristics:

(i) a general ontological innovation, and new guiding picture of nature, in its claim that higher-level working individuals, and all of their activities and working properties, are composed by lower-level working individuals, and their activities and properties;

(ii) a novel family of models/explanations underwritten by this ontological innovation in the integrated compositional models/explanations surveyed in Section 1 (that is, Dynamic, Standing, and Analytic compositional models);

and

(iii) new methodologies, including the strategy of searching for compositional models/explanations, among others.

It is an interesting question, which I explore in coming sections, whether other Global Research Movements have features mirroring some, or all, of characteristics (i)–(iii). But another looming issue is whether such Global Ontological Research Movements are ever successful? In the next subsection, I show that Everyday Reductionism provides an answer to this question as well.

2.3 The Twentieth-Century Success of the Everyday Reductionism

The array of compositional models/explanations supplied by twentieth-century science is overwhelming in number and extent. But let us start by noting how successful compositional models/explanations were provided for the problem

---

[27] For example, Dennett (1978) is basically advocating for a new discipline of cognitive science to pursue compositional explanations given their PEP, since we can then explain properties like thinking using very different properties of parts.

cases in the first reduction–emergence debate. For the discovery of enzymes allowed researchers to compositionally explain digestion. And the rise of quantum mechanical accounts underwrote plausible compositional models for the properties of NaCl, and other substances, using properties of their atomic parts and their still more fundamental constituents.

In addition, the broader potential of the Everyday Reductionist's suite of models, that I just flagged, was also put to work, and compositional models/explanations even made the headlines in popular media. For instance, the discovery of DNA underpinned compositional, and other, explanations of the inheritance of traits by offspring from their parents, and then waves of other compositional models/explanations supplied by molecular biology. But, beyond such headline cases, we should mark the breadth and depth of the more quotidian compositional models/explanations provided by twentieth-century science beyond these headline cases.

The *depth* of the compositional models/explanations provided in the twentieth century was already illustrated in Section 1. We saw how in biological sciences, such as physiology, cell biology, and molecular biology, researchers provided compositional explanations of certain kinds of individual, like a skeletal muscle, but also compositional explanations of the activities of such individuals, like contracting, and compositional models of their properties, such as their mass, energy, and strength. Furthermore, we saw that such integrated compositional models/explanations were offered at a number of levels, too. Together, such integrated compositional models provide a comprehensive compositional understanding of not just certain kinds of individuals, but also their activities and properties, at various levels. And this is true of all the organs of the body (aside from the brain), their tissues, cells, and molecular constituents, too.

In fact, the depth of such compositional models/explanations led scientists to develop connected notions such as that of a "level of organization" or "compositional level." Thus, by the middle of the twentieth century, we find scientists like the organelle researcher Alex Novikoff (1945) advocating for the widespread adoption of such levels used in connection with the coalitions, including compositional models/explanations, that had been developed for the human body in physiology, cell biology, and molecular biology.[28]

We must additionally mark the *breadth* of the compositional models/explanations that were developed, since this goes well beyond the biological disciplines and stretches across to sciences like geology or meteorology, and all the way down to disciplines like chemistry and materials science. For instance, work in the twentieth century allowed us to compositionally explain the movement of the

---

[28] See Gillett (2021) for a reconstruction of the nature of such compositional levels in physiology.

Earth's surface, an activity, using compositional relations to activities of the Earth's parts in continental plates and currents of magma. And, to take an example at a lower level, in the twentieth century we compositionally explained the refractive index of a crystal, a property, using compositional relations to the properties and relations of the crystal's constituent atoms or molecules. And a myriad of other compositional models can be found across various sciences.

2.4 We are All (Everyday) Reductionists Now: The Resolution of the First Set of Reduction–Emergence Debates

By the mid to late twentieth century, scientists took the first set of reduction–emergence debates to have been resolved by the comprehensive provision of such compositional models/explanations for working entities across the sciences. Mainstream scientists now all accept that all the working individuals in nature, and their activities and working properties, are composed by parts, and their activities and working properties, that are ultimately composed by the entities of fundamental physics. We can thus see how a key issue in endogenous metaphysics, over the truth of what philosophers term "physicalism," was plausibly empirically resolved by the provision of successful compositional models – hence illustrating how issues in endogenous metaphysics are often empirically resolvable.[29]

Today, the Nobel Prize–winning physicist Steven Weinberg is therefore right to claim all scientists are now "reductionists" *if* we mean by this that they are all Everyday Reductionists.[30] Let us dig into Weinberg's claim in a little more detail, since it is telling in a couple of ways.

First, Weinberg's statement expresses an important truth if we read "reductionism" as Everyday Reductionism. For researchers across the sciences, at whatever level, now endorse universal composition and are open to the ubiquity, and utility, of compositional models/explanations. Everyone in the sciences is now therefore a "reductionist" in the sense of searching for "reductions," or "reductive," explanations, in the shape of compositional models/explanations for working entities – albeit alongside all their other kinds of model/explanation. Crucially, under such a "reduction" with Everyday Reductionism, namely in a compositional model/explanation, we have at least *two* compositional levels of entities involving both wholes *and* parts. For compositional models/

---

[29] The formulation of "physicalism" in philosophy has been problematic given the neglect of compositional models in the sciences. See Crook and Gillett (2001) for my favored kind of formulation. And the formulation of "physicalism" plausibly matters in philosophical practice. See, for example, Elpidorou and Dove (2018) on what a research program for consciousness looks like under a physicalism framed around the findings of Everyday Reductionism as opposed to standard philosophical formulations of "physicalism."

[30] See Weinberg (1992, p. 62), among other places.

explanations endorse both a whole (or its activities and properties) as explanandum as well as parts (or their activities and properties) as explanans. This "reductionism" thus embraces a *plurality* of entities, compositional levels, and integrated ontic models where its compositional models/explanations *supplement* causal, mechanistic, and still other ontic models/explanations.

Second, Weinberg's remark is also telling because it flags a continuing problem. Philosophers, but also scientists like Weinberg, fail to explicitly distinguish the *other* "reductionist" view in our contemporary debates from the Everyday version. Rather than a claim about Everyday Reductionism, Weinberg apparently means that everyone in the sciences presently endorses another, more radical, "reductionism" of the kind he advocates. I turn to this distinct position, advocated by Weinberg and others, in the next section.

But we can provide still more evidence that everyone in the sciences is now actually just an Everyday Reductionist by noting remarks of one of Weinberg's opponents in the contemporary debates. For we find the prominent contemporary "emergentist" Robert Laughlin, like Weinberg a Nobel Prize–winning physicist, telling us about the main conclusion of his contemporary "emergentist" position:

> One might subtitle this thesis the end of reductionism (the belief that things will necessarily be clarified when they are divided into smaller and smaller component parts), but that would not be quite accurate. All physicists are reductionists at heart, myself included. I do not wish to impugn reductionism so much as to establish its proper place in the grand scheme of things. (Laughlin 2005, p. xv)

Here again we see a struggle to articulate what "reductionism" is in contemporary debates. However, once we explicitly articulate Everyday Reductionism, then we can see why it is not paradoxical for Laughlin to think present-day emergentists are all "reductionists" and are simply seeking to clarify the appropriate scope of that "reductionism." For such emergentists accept *Everyday Reductionism* and the compositional models/explanations it supplies, but then seek to establish what they take to be the proper understanding of this shared commitment.

2.5 Everyday Reductionism as the Background to Contemporary Reduction–Emergence Debates

We have begun to see that our present cycle of reduction–emergence debates plausibly grows out of, and builds upon, the results of the first cycle. Over the next three sections, I will provide detailed support for the latter conclusion. To underpin this work, let me briefly frame *two* ways in which Everyday

Reductionism may underpin our present, and second, cycle of reduction–emergence debates.

First, as this section makes clear, the empirical successes of Everyday Reductionism, in supplying successful compositional models/explanations, are a central part in these new debates as elements of our understanding of nature. In addition, as we shall see, such compositional models/explanations have allowed us to create new investigative techniques that have provided us with novel empirical findings. Our successful causal, mechanistic and other ontic models/explanations are also important, and so too is the huge range of our other evidence, but compositional models/explanations play an especially prominent role in the newest cycle of reduction–emergence debates – perhaps because these were among the fresh new findings of one of the big turns in the Dynamic Cycle in twentieth-century science.

Second, writers in our contemporary set of debates in the sciences appear to have taken Everyday Reductionism, given its success, as an exemplar of what a Global Ontological Research Movement should be like. That is, recent scientific positions appear to have modeled themselves as movements, to a greater or lesser extent, upon Everyday Reductionism *as well as* embracing, and reacting to, its array of successful compositional models/explanations. Subsequent movements may hence have sought (i) to sketch a novel global ontological picture – a new guiding view – involving an ontological innovation. Consequently, such positions may also have defended (iii) new methodologies built around this new guiding picture; and/or (ii) pioneered novel models/ explanations using these ontological innovations. In the coming sections, I also explore this second suggestion about the influence of Everyday Reductionism.

2.6 Philosophically Overlooked but a Towering Success of Twentieth-Century Science

Though overlooked by mainstream philosophy, through its neglect of compositional models/explanations, Everyday Reductionism is in fact the "reductionism" that dominated the sciences for much of the twentieth century and which is now endorsed by all working scientists. It is important for mainstream philosophy to finally explicitly acknowledge this historical fact. For Everyday Reductionism provides not only an example of a Global Ontological Research Movement, but a highly successful one to boot. The search for, and provision of, our suite of integrated compositional models/explanations was one of the highlights of twentieth-century science and is an everyday methodological mainstay of contemporary science.

When we fail to provide theoretical accounts of foundational phenomena in the sciences themselves, then things often ultimately go awry in our understanding of science in one way or another. In the next section, I now highlight how the theoretical failure to explicitly frame Everyday Reductionism, or its underlying compositional models/explanations, has had damaging implications in the sciences as some scientists have flagged.[31] For I show both philosophers *and* scientists have subsequently failed to carefully differentiate Everyday Reductionism from the *other* "reductionism" that Weinberg, and other scientists, have subsequently championed in our new set of reduction–emergence debates.

3 The *Other* Reductionist Movement: Appreciating Fundamentalism

In present-day science, there is another "reductionist" movement, distinct from Everyday Reductionism, in our new reduction–emergence debates. This is what I term "Fundamentalism" and it is defended by prominent scientific adherents like Weinberg and other physicists, biologists like Francis Crick, Richard Dawkins, and E. O. Wilson, and others.[32] Unfortunately, both scientists and philosophers often confuse Everyday Reductionism and Fundamentalism, and refer to each as "reductionism," though we shall see that they have sharply different commitments.

Fundamentalism takes Everyday Reductionism as both its *starting point* but also its *target*. Like the other scientific positions in our present debates, Fundamentalism takes Everyday Reductionism as its starting point by embracing our array of successful compositional models/explanations and advocating our continuing search for such models. But Fundamentalism also takes Everyday Reductionism as its target because it argues that reflection on compositional models/explanation shows we ought to abandon the guiding picture of Everyday Reductionism in favor of a more austere view of nature.

The engine of Fundamentalism is a kind of theoretical argument that, as I outline in Section 3.1, has long attracted philosophers from at least the ancient Buddhists onwards. These are ontological parsimony arguments about a whole and its parts that the Fundamentalist argues establish that in understanding compositional models/explanations we should accept nothing but parts, and collectives of them, rather than parts *and* wholes.

Fundamentalism consequently presses, as I sketch in Section 3.2, a different guiding picture of nature solely involving parts and various scales of collectives

---

[31] Scientists like Ernst Mayr (1988) have suggested that using "reductionism" in this ambiguous way has been damaging to the scientific debates.

[32] Crick (1966), Dawkins (1982, 1987), Wilson (1998), and Weinberg (1992, 2001), among others.

of them – hence radically *subtracting* from Everyday Reductionism's picture of nature by abandoning compositional levels of both parts and wholes. I also flag how Fundamentalism is also implicitly committed to an ontological assumption I dub the Simple view of nature which takes the activities and properties of individuals to be the same across all conditions, including all scales of aggregations and collectives, and hence to be covered by the same simple set of determinative laws in the simplest to the most complex situations.

Filling out the commitments of Fundamentalism, in Section 3.3, I detail how Fundamentalists like Weinberg now defend what they label a "compromising" view taking higher sciences, their laws, explanations, and predicates to be significant and indispensable in a variety of ways.[33] Nonetheless, I illuminate how even compromising Fundamentalists, like Weinberg, still defend a special status for the lower sciences studying components and the determinative entities and laws they illuminate.

I conclude, in Section 3.4, that there is a case that Fundamentalism is a Global Ontological Research Movement, but I show we need to put an asterisk next to this claim. For I also note that Fundamentalism offers researchers no new models/explanations beyond those already pioneered by Everyday Reductionism, so Fundamentalism is not plausibly an iteration in scientific practice of the Dynamic Cycle.

3.1 Ontological Parsimony Arguments and the Engine of Fundamentalism

It is interesting to compare the "engines" of Everyday Reductionism and Fundamentalism. The engine of Everyday Reductionism is obviously the provision of its suite of compositional models/explanations. In contrast, Fundamentalism is driven by a theoretical argument that the Fundamentalist claims we should apply to such compositional models/explanations to draw a surprising conclusion.

Many ordinary people are attracted to the same kind of conclusion once they appreciate folk compositional models/explanations. And such reasoning has long been pressed by a variety of thinkers. For example, we find sophisticated versions of such arguments stretching back at least to the ancient Buddhist philosophers.[34] To get a sense of such reasoning, consider how it proceeds in an ordinary case familiar to the Buddhists such as the folk compositional explanation of a bullock cart. We know the parts of such a cart in boards, axles, wheels,

---

[33] Weinberg (2001, p. 13). For similar sentiments, see Dawkins (1987, pp. 14–15).
[34] See Siderits (2007) for an excellent introduction to the main strands of ancient Buddhist reductionism.

and so on. And we can explain properties and activities of the cart using compositional relations to the properties and activities of these parts.

In this type of example, the Buddhist suggests we have two hypotheses to choose between about the entities that we should take to exist in such a case of compositional explanation. The usual hypothesis takes there to be *both* parts, in boards, axles, wheels, and so on, *and also* a whole, in a bullock cart. In contrast, the Buddhist reductionist's favored hypothesis is that there are the parts *alone*, albeit the interrelated and organized collective of parts that we use to explain the cart and its activities and properties.

The Buddhist has an argument for their favored hypothesis over the usual one. For the Buddhist reductionist claims that our compositional explanations mean that we can explain all the properties and activities of both whole and parts using the parts alone. And the Buddhist applies the Parsimony Principle that when we have two hypotheses that explain equally well, then we should accept the hypothesis that posits fewer entities. Since the two hypotheses putatively explain everything equally well, but the Buddhist reductionist's hypothesis posits fewer entities, the Buddhist concludes that we should accept that in such cases of compositional explanation there are parts alone.

The contemporary strategy of Fundamentalism is to apply this kind of ontological parsimony argument in the sciences to the army of compositional models/explanations that resulted from the success of Everyday Reductionism. Consider an example of such reasoning in what I term the Argument from Composition (Gillett 2007a, 2016a, ch. 3).

As we saw in Section 1, a successful Dynamic or Standing compositional explanation allows one to explain the activities or properties of a whole using the activities or properties of its parts. For example, we saw how we compositionally explain the muscle contracting using the activities of its constituent proteins, but not vice versa. The Fundamentalist therefore first draws the subconclusion that our successful compositional explanations mean that we can explain all the activities and properties of both parts and wholes using the parts alone. And the Fundamentalist claims we again have two hypotheses about the entities any compositional explanation should be taken to be about. On one side, we have the usual hypothesis that we have a whole and its parts (i.e. the proteins plus a muscle). On the other side, we have the Fundamentalist's favored hypothesis that we have parts alone (i.e. the interrelated and organized proteins alone). But, given their subconclusion that we can explain everything with parts (and their activities and properties) alone in cases of compositional explanation, the Fundamentalist claims these hypotheses explain everything equally well. Furthermore, the latter hypothesis is simpler than the former. The Fundamentalist hence applies the Parsimony Principle, along with the crucial subconclusion, to conclude that we should only accept that there are parts, and their activities and properties, in any case of compositional explanation.[35]

Understanding the engine of Fundamentalism, in such parsimony arguments, we can now explore the details of this scientific position. But we should first note related views in recent philosophy. Philosophers have long overlooked compositional models/explanations in the sciences and focused primarily on causal explanations. Against this background, the philosopher Jaegwon Kim (1993a, 1993b) pioneered various ingenious arguments revolving around *causal* relations that get to the same conclusions as the Fundamentalist's ontological parsimony reasoning about *compositional* relations. Philosophical positions advocating versions of an ontological reductionism similar to Kim's, or to Fundamentalism itself, have subsequently been developed in the philosophy of science (Rosenberg 2006), philosophy of mind (Heil 2003), and metaphysics (Van Inwagen 1990; Merricks 2001).

3.2 Appreciating Collectivist Ontology and the Simple View of Nature: A New Guiding Picture

The Fundamentalist's ontological innovation is that nature involves parts alone and this leads to the adoption of a guiding picture of nature involving no wholes, no compositional relations, and no compositional levels. But it needs to be emphasized that Fundamentalism does not endorse a guiding picture that takes nature to be a "dust cloud" of isolated, and unrelated, fundamental parts as its critics often wrongly claim.

Fundamentalism accepts compositional explanations and concludes we should only endorse the entities used as the explanans at the "bottom" of such explanations – that is, all the entities at the bottom (or sometimes top) of the figures in Section 1. Rather than being isolated and unrelated parts, the explanans in such models/explanations are always what I term *collectives* of interrelated and organized parts as we have seen in our scientific cases. For example, we compositionally explain the muscle's contracting using the coordinated activities of many interrelated and organized skeletal muscle cells or proteins.

Fundamentalism thus endorses not only isolated parts in the simplest situations, such as the isolated particles studied using supercolliders, but also many collectives of interrelated parts, of increasing scales, found in atoms, molecules, cells, tissues, and so on, whose behaviors and properties we illuminate using the models/explanations of various sciences. Such collectives are obviously of many different scales, since the collectives of fundamental parts in,

---

respectively, a myosin protein, skeletal muscle, and a complete human body are each massive collectives but of distinct scales. The guiding picture of Fundamentalism is that nature has different scales of complex collectives of parts, rather than the nature of Everyday Reductionism with its compositional levels of parts and wholes.[36]

In addition to this ontologically innovative guiding picture, it is also important to make explicit another ontological assumption of Fundamentalism. This is what I shall term the Simple view of nature that takes all individuals to behave in the same ways, covered by the same determinative laws, under all conditions.[37] To appreciate the reach of the Simple view, consider what it says about individuals as they form relations with each other and aggregate into greater and greater collectives.

First, under the Simple view, the activities and powers of aggregated individuals are *determined only by other entities at the component level or still lower levels*. And second, the determination of these new behaviors and powers of aggregated individuals is *continuous* across both simpler and more complex collectives because it is exhaustively determined, to the degree it is determined, by the same simple set of laws and principles applying to the simplest systems. If we put these points in terms of the laws, or 'principles of composition' that describe aggregation, then the Simple view takes the simple set of laws, or principles of composition, describing all aggregation to refer only to component entities (at the same or still lower levels) and to be continuous, and exhaustive, in nature across all collectives of individuals however simple or complex.

Given the character of the Simple view, Fundamentalism can "dream" of a "Final Theory" as Weinberg (1992) puts it. That is, an account of the laws applying to the most fundamental parts in simple systems that exhaust the determinative laws and principles applying to all such individuals wherever they are found. Since the Fundamentalist takes these fundamental parts to be the only individuals that exist, given their parsimony reasoning, such a simple set of laws would hence truly be a Final Theory – and we can appreciate that the truth of the Simple view is a precondition for such an account.

3.3 Indispensable Higher Sciences and Laws, and New Methodologies under Fundamentalism

In the 1960s, as molecular biology boomed after the discovery of DNA, RNA, and so on, and the provision of a wave of compositional models in biochemistry, Fundamentalists in biology often pressed what Dan Dennett (1996) aptly terms a "greedy" reductionism. This is a position that holds that the lower sciences

---

[36] See Gillett (2016a, ch. 3–4) for further discussion of the resulting picture of nature.
[37] See Gillett (2016a, ch. 3).

studying components can perform *all* explanatory or descriptive tasks, including all the work of higher sciences. For example, molecular biology was widely claimed to be able do all the explanatory work in biology. But this contention that higher sciences are *dispensable* has not been borne out by later science for, as we saw in Section 1 (and as writers like Kitcher (1984) argue in detail), we still need, and use, all manner of models/explanations representing "muscles," "bodies," "cells," and so on.

Consequently, Fundamentalists like Weinberg now defend what they label a "compromising" view taking higher sciences, their laws, explanations, and predicates to be significant and indispensable in a variety of ways.[38] For example, in the case of statistical mechanics, Weinberg contends that:

> The study of statistical mechanics, the behavior of large numbers of particles, and its application in studying matter in general, like condensed matter, crystals, and liquids, is a separate science because when you deal with very large numbers of particles, new phenomena emerge . . . even if you tried the reductionist approach and plotted out the motion of each molecule in a glass of water using the equations of molecular physics . . ., nowhere in the mountain of computer tape you produced would you find the things that interested you about water, things like turbulence, or temperature, or entropy. Each science deals with nature on its own terms because each science finds something in nature that is interesting. (Weinberg 2001, p. 40)

Here we see a commitment to the idea that the various sciences find aspects of nature that are significant and that we hence need a variety of sciences with their own proprietary predicates to capture all the truths about nature. But how can a *Fundamentalist*, with her austere guiding picture, accept that higher sciences are indispensable in this way?

The key point is that the Fundamentalist accepts there are various scales of collectives of parts, each of which together behaves differently than the other scales. The collective of parts in a myosin protein behaves in ways that leave a chair unaffected, whilst the collectives of parts in an elephant behave in ways that break a chair. As Weinberg notes:

> . . . various silly things . . . might be meant by a final theory, as for instance that the discovery of a final theory in physics would mark the end of science. Of course a final theory would not end scientific research, not even pure scientific research, nor even pure research in physics. Wonderful phenomena, from turbulence to thought, will still need explanation whatever final theory is discovered. The discovery of a final theory in physics will not necessarily even help very much in making progress in understanding these phenomena. (Weinberg 1992, p. 18)

---

[38] Weinberg (2001, p. 13). For similar sentiments, see Dawkins (1987, pp. 14–15).

Here Weinberg makes the point that even if lower sciences complete a Final Theory that describes the only determinative entities and the only determinative laws, then Fundamentalism *still* accepts that lower sciences do not supply predicates, explanations, and theories that suffice to express all the truths about nature, including all the true explanations and laws. The deeper idea of this compromising Fundamentalism is that new predicates, and new laws, are needed to capture, and hence express, the truths about the various scales of collectives – and that is why higher sciences and their predicates, and laws, are indispensable.[39]

However, we should immediately note that the subject matters of higher and lower sciences are still not taken to be equal even under such a compromising Fundamentalism. For we need to remember that collectives are not taken to be new individuals, nor are the individuals within collectives taken to obey any new determinative laws. Under Fundamentalism, and its Simple view of nature, the only determinative entities are the parts in such collectives, and the only determinative laws holding of the parts in collectives are still the simple set of laws holding of them in the simplest systems described by a Final Theory. As Weinberg frames the situation:

> Think of the space of scientific principles as being filled with arrows, pointing toward each principle and away from others by which it is explained. These arrows of explanation have already revealed a remarkable pattern: they do not form separate disconnected clumps, representing independent sciences, and they do not wander aimlessly, – rather they are all connected, and if followed backward they all seem to flow from a common starting point. This starting point is what I mean by a final theory. (Weinberg 1992, p. 6)

Under this picture, the laws of higher sciences are not determinative laws; rather they are generalizations that describe the behaviors of collectives that are determined, insofar as they are determined, by the only determinative laws in the simple set of laws holding of the parts in the simplest systems described by the Final Theory of fundamental physics.

Although higher sciences might be taken to be indispensable under compromising Fundamentalism, lower sciences studying parts, and the laws they articulate, are thus still given a special status. And Fundamentalism also results in new methodologies reflecting this underlying point. For example, Fundamentalism is argued to lead to a form of methodological priority in funding allocation that generalizes to all lower sciences. About this methodology, Weinberg tells us that:

---

[39] See Gillett (2016a, ch. 4), for a detailed treatment of such a compromising Fundamentalism. Dennett's famous "Real Patterns" paper (Dennett 1991) can be read as articulating this type of compromising Fundamentalism.

All I have intended to argue for here is that when the various scientists present their credentials for public support, credentials like practical value, spinoff, and so on, there is one special credential of elementary particle physics that should be taken into account and treated with respect, and that is that it deals with nature on a level closer to the source of the arrows of explanation in other areas of physics. But how much do you weight this? That's a matter of taste and judgment. (Weinberg 2001, p. 23)

Here we see Weinberg acknowledging the range of factors, and difficulty, in funding decisions. Nonetheless, we also see the contention that Fundamentalism underwrites the methodology of giving the sciences studying parts a special claim to funding priority over sciences studying composed entities. And we find related methodological implications of Fundamentalism, given its guiding picture and underlying assumptions, for a range of other practices beyond funding allocation.

3.4 Fundamentalism as a Global Ontological Research Movement with an Asterisk

The work of this section shows that the rise of Fundamentalism in the sciences in the late twentieth century is a manifestation of the iterative nature of these debates. Although Fundamentalism is a minority view in the sciences in the early twentieth century, defended by a few scientists like Eddington as I suggested in Section 1, it is no accident that Fundamentalism is a more widely espoused position by the late twentieth century. For the successful provision of compositional models/explanations by the Everyday Reductionism movement, across the twentieth century, resolved the first set of reduction–emergence debates and provides the fuel for new views such as Fundamentalism with its ontological parsimony arguments about compositional models.

Furthermore, the work of this section also confirms that Fundamentalism is a position distinct from Everyday Reductionism. Fundamentalism has a different ontological innovation, guiding picture of nature, and methodologies than Everyday Reductionism. The ontological innovation of Fundamentalism is that there are parts alone challenging the core commitment of Everyday Reductionism to wholes and parts. While Everyday Reductionism accepts there are levels of parts and wholes, and a plural array of determinative entities, laws, and explanations across many sciences, Fundamentalism rejects all of these claims. We can thus confirm that these are very different positions and that going forward we finally need to carefully distinguish, on any occasion, which position we are referring to in the sciences when we talk about "reductionism."

Is Fundamentalism another Global Ontological Research Movement? On one side, the positive case we can now make is that Fundamentalism mirrors feature (i), which we found in Everyday Reductionism, in pressing an ontologically innovative view of nature as solely consisting in parts and collectives of them. And Fundamentalism also has an analog of characteristic (iii) of Everyday Reductionism in its novel methodologies such as prioritizing for funding those sciences that study parts, amongst others. There is thus some positive support for taking Fundamentalism to be a Global Ontological Research Movement.

However, on the other side, there are also concerns that arise when we consider whether Fundamentalism is pursuing the Dynamic Cycle. For is Fundamentalism a position offering ontological innovations to produce new ontic models/explanations to solve scientific problems? Answering this question is much more fraught and in interesting ways.

As we have seen, the engine of Fundamentalism is a kind of theoretical argument, in ontological parsimony reasoning, like the Argument from Composition, about the compositional models/explanations pioneered by Everyday Reductionism. The ontological innovation of Fundamentalism – in a nature of parts and collectives of them alone – grows from these theoretical arguments and results in a putatively better interpretation of the *same* models/ explanations that Everyday Reductionism pursues. We can thus see a plausible case that Fundamentalism *offers researchers no new models/explanations* to use in their research and thus lacks feature (ii). For Fundamentalism is simply focused on reinterpreting the nature of the entities described by the models/ explanations pioneered by Everyday Reductionism.

And there is a follow-on point. Fundamentalism is thus not primarily focused on offering new models to solve ongoing problems in scientific research, since its focus is on using theoretical arguments to draw conclusions about the import of our existing models/explanations. Consequently, it appears that Fundamentalism is not pursuing the Dynamic Cycle.

Depending upon how closely one wants to associate Global Ontological Research Movements in the sciences with the Dynamic Cycle, and the search for new models, one may thus have real concerns about whether Fundamentalism is such a movement after all. Given the points in support of Fundamentalism as a Global Ontological Research Movement, since it shares features (i) and (iii) with Everyday Reductionism, I will continue to treat it as an example of such a movement. But I suggest we should place an asterisk by this claim, given the fact that Fundamentalism is not pursuing the Dynamic Cycle and lacks feature (ii) because it is not offering new models/explanations to researchers.

The latter points also suggest a diagnosis about why Fundamentalism has found fewer adherents in contemporary science than its "emergentist" rivals. For example, in typically colorful fashion, Richard Dawkins, a prominent exponent of Fundamentalism, complains that in the sciences "Reductionism is a dirty word, and a kind of 'holistier than thou' self-righteousness has become fashionable" (Dawkins 1982, p. 113). First, we have confirmed that this antipathy is targeted on Fundamentalism, rather than the Everyday Reductionism that all sides endorse. And second, the failure to offer researchers new models/explanations may explain why they have not adopted Fundamentalism in larger numbers. If opposing "emergentist" views do provide researchers with such new resources, then we have a clear reason why more researchers might favor such "emergentist" approaches in practice.

3.5 The Other Reductionism as an Outlier in the Sciences?

In his popular exposition of Fundamentalism, we saw earlier that Weinberg (1992) famously declared about his fellow scientists that "we are all reductionists now." The work of this section confirms that this statement is dangerously ambiguous. All contemporary scientists are plausibly *Everyday Reductionists* now – for they all accept a plethora of compositional models/explanations and the utility of searching for such models/explanations across nature. But Weinberg's declaration concerns the *Fundamentalism* that he and others have championed in the second cycle of reduction–emergence debates and it is far from clear that most scientists presently endorse this *other* "reductionism."

We have therefore confirmed the importance of disambiguating what one means by "reductionism," in both science and philosophy. Everyday Reductionism and Fundamentalism have radically different guiding pictures of nature. Furthermore, Everyday Reductionism focused upon ongoing problems in the sciences that it offered new models/explanations to address, while Fundamentalism apparently offers no new models/explanations to help with such problems.

An irony hence also looms. Fundamentalism is one of the few positions in recent reduction–emergence debates in the sciences that philosophers have in fact recognized to some degree. For Fundamentalism mirrors, in many ways, the versions of ontological reductionism that I noted earlier have recently been defended across various areas in philosophy by writers like Kim, Heil, Merricks, Rosenberg, Van Inwagen, and others. But Fundamentalism may be the odd one out in the scientific debates for the same reasons it is a position familiar to philosophers. Fundamentalism in the sciences is not an iteration of the Dynamic Cycle, nor plausibly an example of endogenous metaphysics, since it does not press a global ontological innovation to underwrite new models/explanations to address ongoing scientific problems. Instead, Fundamentalism pursues the kind of metaphysics that philosophers engage in within their debates where ontological parsimony figures so centrally because this purely theoretical work is not intended to result in new scientific models to be applied to nature itself and hence evaluated using the resulting empirical findings.

Yet Fundamentalism is the outlier in our present round of reduction–emergence debates in the sciences *only if* opposing "emergentist" and other positions do offer new resources to address ongoing scientific problems. And discerning whether "emergentist" views have such an orientation requires that we have some sense of the kind of problem currently faced in the sciences. In the next section, I therefore consider an ongoing difficulty in many sciences that grows out of the successes of Everyday Reductionism and which has been the focus of many present-day "emergentists" in the sciences.

4 Challenging Compositional Cases: Ongoing Problems, and Responses, in Today's Science

Successful new models/explanations often beget novel experimental and other investigative techniques. And after the explosion of compositional models/ explanations in the twentieth century, investigative techniques, partly built using these models, finally provided researchers with quantitative accounts of the behaviors of the parts within many complex wholes. The result includes what I term Challenging Compositional Cases in examples of wholes about which we now have coalitions of models, including successful compositional models along with causal, mechanistic, and other models. But where we also now have quantitative evidence about the activities of the parts within these wholes. In many of these examples, I show that we face an ongoing scientific problem of understanding the behavior of the parts in the relevant wholes.

To illustrate the nature of Challenging Compositional Cases, in Section 4.1, I briefly consider a couple of concrete examples to give a sense of their characteristics. I look at the arguments of Robert Laughlin about the behavior of electrons in superconductors. And I examine the work of a team of systems biologists and philosophers of science on the behavior of proteins in cells. I also note a range of other examples with apparently similar features across many sciences and levels of nature.

Using the latter examples, in Section 4.2, I draw out some common features of Challenging Compositional Cases. I highlight how in these examples we have successful coalitions of integrated models and new quantitative empirical findings about the behavior of parts in the relevant wholes. But I then highlight how this evidence itself implies that we still do not understand the behaviors of the parts in the relevant wholes – hence leaving us with an ongoing problem in contemporary science. In the face of such difficulties, in Section 4.3, I examine whether approaches we have examined earlier are adequate to handling Challenging Compositional Cases. I suggest there are reasons to conclude they are not – hence leaving us with a scientific difficulty.

One response to the problem in Challenging Compositional Cases is to pursue new iterations of the Dynamic Cycle. I therefore highlight, in Section 4.4, one global ontological innovation that theorists, across the sciences and philosophy, have championed to this end. This is what I term the Conditioned view of nature and its key contention is that parts really do behave differently in wholes – hence such parts are taken to have what I term "differential" activities/behaviors, and powers, that are different from what they would have if the laws and frameworks adequate to such parts in simpler systems were exhaustive in the relevant whole.

Perhaps most importantly, in Section 4.5, I highlight how this ontological innovation alone, or in combination with another innovation, can drive a pair of positions that each underwrites its own new family of models/explanations. I finish the section by exploring one of these positions in detail, what I term the Causally Conditioned view, and the new family of models/explanations it underwrites in Challenging Compositional Cases.

4.1 Exploring a Common Kind of Example in Contemporary Science

In this subsection, I briefly look at a couple of examples of Challenging Compositional Cases, in Sections 4.1.1 and 4.1.2, one from a physical science and the other from a biological discipline. In Section 4.1.3, I then note the wide range of other putative examples of Challenging Compositional Cases from the activity of molecules to the flocking behavior of birds.

4.1.1 Electrons in Superconductors: Challenging Compositional Case (I)

Let us start with the arguments made by the physicist Robert Laughlin about the constituent electrons in superconductors (Laughlin 2005). Laughlin's claims center around our quantitative knowledge about the actual behaviors of electrons in superconductors as well as what the quantum mechanical laws holding of electrons in simpler systems imply about their behavior in superconductors.

The physicist turned philosopher of science Sunny Auyang provides an accessible account of high-energy superconductivity articulating the key features of the case that Laughlin uses in his arguments. Auyang tells us that in high-energy superconductors:

> The superconducting current cannot be obtained by aggregating the contributions of individual electrons; similarly it cannot be destroyed by deflecting individual electrons. The microscopic mechanism ... baffled physicists for many years, and was finally explained in 1957 by the Nobel prize-winning BCS theory developed by John Bardeen, Leon Cooper, and Robert Schrieffer.

An electron momentarily deforms the crystal lattice by pulling the ions towards it, creating in its wake a surplus of positive charges, which attract a second electron. Thus the two electrons form a pair, called the Cooper pair ... All the Cooper pairs interlock to form a pattern spanning the whole superconductor. Since the interlocking prevents the electrons from being individually scattered, the whole pattern of electrons does not decay but persists forever in moving coherently as a unit. The structure of the inter-locked pattern as a whole, not the motion of the individual electrons, is the crux of superconductivity.

... No special ingredient is needed for [this] emergence; superconductors involve the same old electrons organized in different ways. The structured organization in superconductivity constrains the scattering of individual electrons and forces them to move with the whole, which can be viewed as a kind of downward causation. (Auyang 1998, pp. 180–81)

Here every entity in the case at the higher level is fully composed and we have accounts of the relevant components underpinning compositional models/ explanations alongside an array of other successful, integrated models/explan-ations. Furthermore, we have a precise quantitative understanding of the behav-iors of electrons in superconductors.

These features of the example are eventually used by Laughlin to argue that with high-energy superconductivity we have an interesting kind of "emergence" and downward whole-to-part determination. I explore those wider claims in the next section. But these wider claims rest on narrower conclusions about the behavior of the parts, namely electrons, in this case. So let me first focus on Laughlin's arguments about those features of the case.

Laughlin (2005) claims that we know the quantum mechanical laws holding of the component electrons in simpler collectives and hence the laws that exhaustively determine, to the degree that they are determined, the components in those simpler collectives. We have also just seen that we have compositional models/explanations of both superconductors and these simpler systems. Lastly, we now have precise quantitative accounts of the behaviors of the component electrons in such superconductors.

Putting the latter together, Laughlin's narrower argument is that we know that the quantum mechanical laws holding in simpler collectives would determine the component electrons have *different* behaviors than we find in electrons in superconductors using our recent investigative techniques. In superconductors, Laughlin concludes, we thus know that *parts behave differently*. And it is important to note the special nature of these different behaviors.

Individuals that are parts often behave differently in wholes, but nonetheless the behaviors are still determined by the laws and frameworks in simpler systems. However, in this case the difference in behavior is of a distinct kind.

Laughlin concludes that we know the parts have different activities/behaviors (and hence powers) than they would have if the laws and frameworks that apply in simpler systems were exhaustive. And it is only this very particular kind of behavior that I am terming a *differential* activity.

As I noted earlier, Laughlin is ultimately like other scientists in defending a richer position about "emergence," and "downward" determination, in order to make sense of such cases in the broader "emergentist" position I survey in Section 5. But this narrower conclusion is, on its own, a striking one. And it becomes more intriguing when we see similar conclusions drawn about other examples.

4.1.2 Proteins in Eukaryotic Cells: Challenging Compositional Case (II)

To vary our diet, let us move to biology to consider the behaviors of the proteins that compose a eukaryotic cell, discussed in detail by an interdisciplinary research team consisting of philosophers of science, in Robert Richardson and Achim Stephan, as well as prominent systems biologists, in Fred Boogerd, Frank Bruggeman, and Hans Westerhoff (Boogerd et al. 2005).

Boogerd et al. look at examples of the biochemical networks involving the molecular parts, like proteins, that we find in eukaryotic cells and in the simpler collectives forming naturally in bacteria or produced artificially using laboratory techniques like knockout mutants. As Boogerd et al. highlight, the biochemical networks of the parts in eukaryotic cells are formidably complex, involve all manner of feedback loops, and are often only describable using nonlinear dynamics.

Once again, in such examples we now have successful coalitions of causal, mechanistic, and compositional models which describe and explain the properties, and behaviors, of such proteins in simple systems whether in vitro or elsewhere. It is important to emphasize that we have fairly comprehensive, and very successful, compositional models/explanations of eukaryotic cells at the molecular level. And we hence know everything at the cellular level is plausibly composed. In addition, as Boogerd et al. highlight, we have now also collected quantitative evidence about the activities (and hence powers/properties) of such proteins when they are parts of such cells.

Boogerd et al. consequently argue that we can now see that the proteins in cells *behave differently* than they would if the accounts holding of such proteins in simpler systems were exhaustive in these wholes. Like Laughlin, Boogerd et al. consequently defend wider claims about "emergence" of a kind I explore in more detail in the next section. But, for my purposes in this section, Boogerd et al.'s pertinent conclusion is the narrower one – namely, that the proteins in eukaryotic cells also have what I am terming differential activities/behaviors. That is, once more, it is claimed that we have good reason to conclude that the behavior of the relevant proteins is different from the activities they would engage in if the accounts holding of such proteins in simpler systems were exhaustive in the cell.

4.1.3 Other Potential Examples of Challenging Compositional Cases

Scientists point to a wide range of other cases as having similar features to the two examples we have briefly examined – that is, cases where parts appear to behave differently in the strong sense of having differential activities and powers. These examples span the full range of sciences and levels of nature. For instance, researchers have claimed the following cases have the same character:

- Components in thermodynamical systems (Prigogine 1968);
- Atoms and molecules in the liquids involved in the reactions in Benard cells in chemistry (Nicolis and Prigogine 1989);
- Components in slime mold (Garfinkel 1987; Nicolis and Prigogine 1989);
- Neurons in neural populations (Freeman 2000a, 2000b);
- Eusocial insects, like ants and bees, in colonies (Wilson and Holldobler 1988; Mitchell 2012);
- Vertebrates, like birds, in flocks (Couzin and Krause 2003).

And there are still further cases, discussed across various disciplines, also broached as having similar features.[40]

What is so special about these Challenging Compositional Cases? Why might someone think they pose any kind of difficulty? And why might scientists think we need ontological innovations – including categorial or global ones – in order to move research forward in such cases? I now turn to these, and other, questions.

4.2 An Exemplar Problem in Contemporary Science: Challenging Compositional Cases and their Features

As I noted earlier, a lesson of the literature on scientific models is that making complex phenomena cognitively tractable is a challenge. And we need to appreciate that *science itself* is often now a complex phenomenon posing similar challenges. The two concrete examples we briefly looked at, in the previous subsection, illustrate how this is true of Challenging Compositional Cases where we have complex coalitions of successful models/explanations as well as other important kinds of evidence.

To make the scientific situation in such examples more cognitively tractable, I therefore begin this subsection by giving a very general account of the features

---

[40] See Scott (2007) for a still more comprehensive list of such cases.

of the scientific situation that we find in Challenging Compositional Cases. I then highlight why our evidence suggests we are left with an ongoing scientific problem in these examples.

In Challenging Compositional Cases, as we saw with superconductors, and proteins in cells, we have well-confirmed compositional explanations/models of both the relevant wholes and also simpler systems. For example, we have successful compositional models of superconductors and also the simpler systems involving electrons. Similarly, we have successful compositional models of eukaryotic cells and simpler systems such as Petri dishes containing proteins that are parts of such cells.

Furthermore, in Challenging Compositional Cases we have successful causal, mechanistic, and other models about the simpler systems that the parts are also involved in. Laughlin thus notes that quantum mechanical accounts plausibly work well for the behavior of electrons in simpler systems. And we have equally good accounts of proteins and their behaviors in simple systems like that in a Petri dish. In addition, we have plausible compositional models/ explanations of these simpler systems.

As I highlighted in starting the section, the advent of new investigative techniques, often underpinned in part by the advent of the compositional models/explanations, delivered by Everyday Reductionism, has now provided us with quantitative accounts of the activities of parts in the complex wholes in Challenging Compositional Cases as well as in simpler systems. For example, we can now measure the behavior of electrons in superconductors, or proteins in cells, in ways that we previously could not.

Combining the latter models and our new quantitative empirical evidence about parts, we saw that in Challenging Compositional Cases scientists can now make a precise comparison. Researchers have successful frameworks for the behavior of the individuals which are parts in simpler systems and those frameworks make predictions about the behavior of these parts within the complex whole. But the kicker is that in Challenging Compositional Cases our quantitative findings show that the activities of the parts in the relevant wholes are *different* from what they would be if the frameworks and laws for the simpler systems were exhaustive.

Let me summarize these features of the situation in Challenging Compositional Cases where at least the following conditions usually hold for these cases where 'S_1–S_n' are the relevant parts and 'W' is the whole:

(I) for the parts S_1–S_n, we have successful coalitions of integrated causal, mechanistic, and other models/explanations of (and their activities and properties), and/or the laws about them, in simpler systems;

(II) for the whole W, we have successful coalitions of integrated causal, mechanistic, and other models/explanations and/or the laws about them;

(III) for the whole W, we have successful compositional models/explanations of W in terms of the parts S_1–S_n;

(IV) (a) for the parts S_1–S_n, we have quantitative evidence about their behaviors in simpler systems than W which fit with our successful accounts of these parts in these simpler systems; and (b) for the parts S_1–S_n, we now possess quantitative evidence about the behavior of the parts S_1–S_n within the whole W;

And:

(V) the behavior/activity of the parts S_1–S_n in the whole W, framed in (IV-b), is different than it would be if the accounts of these parts in simpler systems, framed in (I), were exhaustive in the whole W.

As this set of conditions highlights, the scientific situation in Challenging Compositional Cases is complex. But the commitments explicitly framed in (I)–(V) highlight the ongoing scientific problem in such examples.

Rather than "gaps" in our knowledge, it is precisely our quantitative empirical knowledge of the activities of parts in both simple and complex situations, alongside our successful models/explanations of these activities in simpler systems, and our successful compositional models/explanations, which frame this difficulty. Our present evidence implies that in Challenging Compositional Cases the behavior of the parts in the relevant wholes still *cry out for explanation and/or understanding*. As writers like Laughlin seek to emphasize, our very precise measurements show that the parts in these wholes – whether electrons in superconductors, or proteins in cells – behave in different ways than our extant accounts of those parts, developed for simpler systems, say that they will behave. Hence such differential activities of parts in these wholes are yet to be explained or understood.

It is important to note this problem does not support returning to the Ontological emergentist positions of the last set of reduction–emergence debate. Challenging Compositional Cases do not involve any uncomposed entities, since we have comprehensive, and successful, compositional models/explanations about the relevant wholes and simpler systems. Instead, Challenging Compositional Cases raise a new difficulty in contemporary science.

4.3 Do We Really have a Problem? The Inadequacy of Extant Responses

Do Challenging Compositional Cases really pose a problem? To answer that question, we first need to assess whether the existing approaches we have considered offer aid. Everyday Reductionism has been immensely successful, so one move would be to apply its methodologies in Challenging Compositional Cases and see whether those help. Unfortunately, as feature (III) of such examples framed, in Challenging Compositional Cases we *already have* successful compositional models/explanations. Consequently, we have exhausted the strategies of Everyday Reductionism in Challenging Compositional Cases.

Challenging Compositional Cases also confirm the limitations of Fundamentalism noted in the previous section. Fundamentalism offers no new models/explanations beyond those of Everyday Reductionism. If the latter is stymied by Challenging Compositional Cases, then Fundamentalism is as well. And, in fact, the commitments of Fundamentalism force it into a narrower response to Challenging Compositional Cases than Everyday Reductionism.

Mark that an Everyday Reductionist can accept the existence of differential activities/powers of parts in wholes. Such differential behaviors do not conflict with the compositional models/explanations advocated by Everyday Reductionism. For one can still use differential activities of parts to compositionally explain activities of wholes. In contrast, the very existence of differential activities of parts conflicts with the deeper commitments of standard forms of Fundamentalism. Let me explain.

Recall that Fundamentalism embraces the Simple view of nature under which the behavior of parts is everywhere determined, to the degree that they are determined, by the same simple set of laws applying in the simplest systems. Under this picture, it is in principle the case that a part can always be explained and understood using the frameworks and laws for the simplest of systems. But a differential activity of a part is precisely a behavior that is different from the behavior that would exist if the frameworks and laws for simpler systems were exhaustive.

The core commitments of the Fundamentalist are thus incompatible with the very existence of the differential activities/powers of parts. Consequently, many Fundamentalists have argued that examples like Challenging Compositional Cases only involve "weak" emergence where *in practice*, but not in principle, there are epistemic reasons preventing us from explaining and understanding the behavior of the parts in such cases using the laws and frameworks that work for simpler systems.[41]

Fundamentalism is therefore forced into the strategy of showing we only have the *appearance* of a difficulty in Challenging Compositional Cases. The latter is a very bold approach, since we have seen that we have putative evidence for differential activities across a wide array of sciences from physics to biology and beyond. The Fundamentalist must commit herself to all these cases being

---

[41] For the classic account of the nature of such "weak" emergence see Bedau (1997).

only apparently problematic. One may reasonably diverge from such a bold assessment, so it is unsurprising that other researchers have thought that a different approach is needed.

4.4 The Conditioned View of Nature as an Ontological Innovation

The sciences appear to face an ongoing problem in Challenging Compositional Cases. There are all manner of potential responses to this difficulty and I cannot consider all of them here. Instead, in the remainder of this section, and in the next section, my focus is solely on researchers who have recently sought global ontological innovations to address Challenging Compositional Cases. One such global ontological innovation is shared by two positions, each of which has subsequently offered a new family of models/explanations to address Challenging Compositional Cases. In this subsection, I therefore start by sketching this shared ontological innovation that I term the Conditioned view of nature.

At its core, the innovation of the Conditioned view is to embrace differential activities of individuals – it allows that some individual engages in one kind of activity under one condition "$", but under a distinct condition "$*" the individual engages in a differential activity – that is, an activity under $* that is different from the activity it would engage in if the laws, models, frameworks, and so forth that hold in $ were exhaustive. One example of such activities is the differential behaviors of aggregated parts in the complex wholes that I consider in the next paragraph. But the Conditioned view potentially applies to individuals whether they are aggregated or not, and hence whether they are a part or not. For example, under the Conditioned view one might thus take an unaggregated individual to have a differential activity under some condition such as causally interacting in some group of individuals.

However, given our focus on Challenging Compositional Cases, it is important to more carefully draw out what the Conditioned view implies about the aggregation of individuals resulting in the organized groups taken to compose wholes. For the Conditioned view allows that when certain parts aggregate, under some condition, then they may have differential activities and powers. That is, when parts aggregate under certain conditions, then the Conditioned view allows these parts to behave in ways different from the ways they would if the laws or principles in simpler systems were exhaustive. Consequently, under those conditions, new frameworks – whether models, explanations, and/or laws – are also needed to explain and understand the behavior of the relevant individuals under those conditions. Furthermore, the Fundamentalist's Simple view of nature, and hence aggregation, is false, since aggregation is not covered under all conditions by the same laws and/or principles of composition.

A novel picture of determinative laws thus goes along with the Conditioned view. There are several options here that I have explored in more detail elsewhere (Gillett 2016a, ch. 7), so let me briefly illuminate the basic points. Under the Conditioned view, individuals, whether parts or otherwise, behave differently under certain conditions and this is not covered by the laws holding in simpler systems. Hence further determinative laws hold of the individual, and its behavior, properties, and powers, under the relevant conditions, where these further laws are not simply derivative from the laws in simpler systems. The Conditioned view thus entails that the fundamental laws of nature are not exhausted by a simple set applying in the simplest system, but also comprise other fundamental laws covering the differential activities of individuals under specific conditions potentially including being a part of a certain kind of whole.

This picture of laws under the Conditioned view again flies contrary to the Simple view of nature assumed by the Fundamentalist. But it is worth noting that the Conditioned view of individuals and their activities, their aggregation, and the laws covering them, does *not* conflict with Everyday Reductionism. One will still be able to provide compositional models/explanations when the Conditioned view of nature applies, but the compositional models/explanations one successfully provides under one condition may be discontinuous with those under another condition because the parts may have differential activities under one, or both, of these conditions.

4.5 New Positions and Models: Surveying our Options in Another Iteration of the Dynamic Cycle

Recent writers appear to adopt the Conditioned view in pursuing an iteration of the Dynamic Cycle to underwrite new models/explanations offering resources in Challenging Compositional Cases. In this subsection, I sketch two such positions that use the Conditioned view in this manner. In Section 4.5.1, I start by looking in detail at what I call the Causally Conditioned position that solely uses the Conditioned view of nature as its ontological innovation. I then briefly outline a second "emergentist" position that I dub Mutualism, in Section 4.5.2, that accepts the Conditioned view of nature *alongside a further ontological innovation* in a form of downward determination, but I defer detailed discussion of this view until Section 5. I finish the section, in 4.5.3, by outlining how the Conditioned view of nature can also underpin a new form of Fundamentalism and some other neglected options.

4.5.1 The Causally Conditioned View, Differential Activities, and Causal Determination: A First Position and Family of New Models

The first position I want to consider is solely built around the Conditioned view of nature as its ontological innovation. This view consequently embraces the existence of differential activities of individuals under certain conditions. In addition, the view assumes that such differential activities of individuals are simply determined, and hence explained, by causal (or at least diachronic) interactions with other individuals. I thus term this the Causally Conditioned view since differential activities/powers are causally determined.

Under the banner of "transformational emergence," philosophers of science like Gil Santos (2015), Paul Humphreys (2016), Alexandre Guay and Olivier Sartenaer (2016), and others, have plausibly endorsed this combination of the Conditioned view and the causal, or at least diachronic, determination of differential activities.[42] This is also a charitable interpretation of the sophisticated, and underappreciated, earlier position of the philosopher of science Miriam Thalos (2013) who eschews composition, reduction and emergence, but endorses something like differential activities and hence the Conditioned view. I also outlined a position of this type in earlier work (Gillett 2016a, ch. 7 and 8) that I labeled Conditioned Fundamentalism, endorsing the differential activities of individuals determined causally by other entities at the same level.

What is most interesting for our purposes here is that all these versions of Causally Conditioned views can underwrite a new family of models/explanations focused on differential activities of individuals. These are what I will term Causally Conditioned models/explanations because they are backed by causal/diachronic relations that determine whether an individual engages in a differential activity (and/or has a differential power). Such models, and the underlying view, can be applied to understand individuals that are not parts, but for our purposes its most obvious application is to the parts in Challenging Compositional Cases.

To this end, consider a diachronic Causally Conditioned model that represents an individual under one condition, at an earlier time, and then under another condition, at a later time, where the individual has some differential activity at the later time. Furthermore, over this period of time, the model posits a causal relation of this individual to some other individual, at the same level, that causes the individual to have its distinctive differential behavior/power. The "condition" where we find the differential behavior may vary, but such models

---

[42] For reasons of space, I leave aside the differences between these writers, and the following ones, and some of the nuances of their views.

all posit a causal/diachronic relation to other individuals as determining, and hence explaining, the relevant differential activity.

Such Causally Conditioned models offer new resources in Challenging Compositional Cases. For one can potentially explain and understand the differential activity of a part, and its effects, using the causal/diachronic relations to other individuals, in the collective that composes the relevant whole, in a Causally Conditioned model. Using Causally Conditioned models, one can therefore seek to explain differential behaviors solely at the same level at which one finds such parts in contrast to the alternative kind of model and position I consider next. Most importantly for our purposes, the Causally Conditioned view hence provides fresh resources to address our ongoing scientific problem.

4.5.2 Mutualism, Differential Activities, and Downward Determination: A Second Position and Family of New Models

A more complex position embraces the Conditioned view of nature alongside a second, categorial ontological innovation. This second innovation is a new kind of whole-to-part determinative relation, at a time, through which a whole determines, and hence explains, a part having certain differential activities/powers. This is what I term Mutualism because we have mutually determinative composed and component entities. This type of emergentist position hence claims that in Challenging Compositional Cases wholes, or their activities or properties, are what determine that parts have differential activities.

This kind of Mutualist position, under various labels, has been defended by scientists who I examine in detail in Section 5. But it has also been espoused by a number of philosophers. Cutting across the differing terminology, and again ignoring various nuances and differences between their views, we plausibly find such Mutualist accounts defended by philosophers including Alicia Juarrero (1999), Robin Hendry (2010), Sandra Mitchell (2012), Alvaro Moreno and Matteo Mossio (2015), and Jason Winning and William Bechtel (2018), among others. I also articulated, and defended the coherence of, Mutualist positions in a series of papers and a book (Gillett 2006, 2011, 2016a, ch. 6 and 7, among others).

What is important here is that these Mutualist positions can again use their ontological innovations to offer us a new family of Mutualist models/explanations, but this time positing *both* differential powers/activities of parts *and* a kind of whole-to-part determination (or an associated kind of downward causation as I outline in Section 5.1). And such models can be used in Challenging Compositional Cases, as I explore in Section 5.3, to potentially allow us to explain and understand the differential behaviors of parts.

We should carefully mark that Mutualist and Causally Conditioned models offer rival, and conflicting, models of the *same* phenomenon involving a differential activity/power of an individual. For Mutualist models take the relevant differential power/activity to be synchronically determined by a composed entity, while Causally Conditioned models take this differential power/activity to be causally (or otherwise diachronically) determined by entities at the same level. However, in the next subsection I note how such models might each be successfully used about *different* phenomena.

This kind of Mutualist position is plausibly endorsed by the most prominent brand of contemporary scientific emergentism, so I will explore it in more detail in Section 5 where I look at the explicit claims of the scientists endorsing this position. But let me first conclude this subsection by considering some other options.[43]

4.5.3 A Different Version of Fundamentalism? Combining Options?

As I earlier mentioned, one can combine the Conditioned view of nature with a novel form of Fundamentalism, what I termed Conditioned Fundamentalism (Gillett 2016a), that can also underwrite Causally Conditioned models. Conditioned Fundamentalism uses ontological parsimony arguments to conclude there are parts and collectives of them alone – hence offering a brand of Fundamentalism. But this view abandons the Simple view, and hence also renounces "dreams" of a Final Theory, to embrace differential activities of individuals and a more complicated set of fundamental laws. I wanted to flag this option for Fundamentalists, since they can consequently continue to embrace the core ideas of Fundamentalism while accepting the innovative Conditioned view – and also offer new models to help in Challenging Compositional Cases.

Given our focus on models, I want to emphasize that it should now be clear that a scientist can offer Causally Conditioned models, and seek to understand the behaviors in Challenging Compositional Cases, *without also accepting* parsimony arguments or the conclusion that there are parts alone. Researchers can thus champion Causally Conditioned models, and the underlying ontological innovation in the Conditioned view of nature that underwrites them, *without* being any kind of Fundamentalist, including this Conditioned variety.

---

[43] Let me note that I take the "contextual" emergentism recently advocated by Bishop, Silberstein, and Pexton (2022) to also plausibly endorse the Conditioned view of nature. But I am unable to say whether this position is a variant of the Causally Conditioned view, Mutualism, or some further option.

Let me also briefly note that researchers can offer each of the distinct kinds of model I have now sketched about different examples. Thus, a researcher might offer Causally Conditioned models in application to one case, whilst applying Mutualist models in a different case. And one might take either version of Fundamentalism to apply in other examples, too. We should be alive to this still more pluralist possibility. For when we begin to apply the various positions, and their models, to the wide array of entities we find in Challenging Compositional Cases, from electrons in superconductors to the birds in flocks, then we may simply be forced to accept the models of distinct positions about different examples. For simplicity is not the sole criterion, as it is in purely theoretical debates, and accommodating our empirical findings across various examples may force us to accept a more pluralist array of models.

4.6 Contemporary Reduction-Emergence Debates as Iterations of the Dynamic Cycle?

The work of this section is not intended to highlight the most important problem in contemporary science. There are obviously many more ongoing scientific difficulties than the kind of problem highlighted here, and I make no claim to be able to rank their relative importance. Furthermore, I have not sought to examine every response to Challenging Compositional Cases, since I have focused solely on reactions based around global ontological innovations. Despite these limitations, the work of this section has taught us some important lessons.

We have confirmed that our contemporary reduction–emergence debates are simply not the debates of the last century. Our present debates do not concern whether everything is composed, as the earlier set of debates did, since we have comprehensive compositional models/explanations in the relevant examples. Instead, our present difficulties arise from our compositional and other models/explanations along with the new quantitative evidence about the behavior of parts within wholes. For the latter together show us that in Challenging Compositional Cases the behaviors of parts in wholes, like superconductors or cells, still cry out for explanation and understanding.

In the final section, I now turn to a more detailed examination of the most prominent form of scientific emergentism in contemporary science which defends the Mutualist view just outlined. That is, a position seeking to under-write a new family of models by combining the Conditioned view with *another* categorial ontological innovation in a new kind of downward, whole-to-part determination.

5 Scientific Emergentism and Its Mutualist Revolution

> The world we actually inhabit, as opposed to the happy world of modern scientific mythology, is filled with wonderful and important things we have not yet seen because we have not looked ... The great power of science is its ability, through brutal objectivity, to reveal to us truth we did not anticipate.
>
> (Laughlin 2005, p. xvi)

The guiding picture of nature to which we subscribe – what we take the ontological structure of nature to be in a broad sense – configures not just what kinds of scientific models/explanations we offer, but also what phenomena we even recognize in nature to start with. In our opening passage, Robert Laughlin, echoing other scientific emergentists like the Nobel-prize winning chemist Ilya Prigogine, claims that we have routinely overlooked all manner of phenomena that did not fit the Fundamentalist's guiding picture of nature which is claimed to be a misleading "myth."[44] Furthermore, emergentists like Laughlin and Prigogine offer categorial ontological innovations, and a new guiding picture, that not only allow us to finally see many natural phenomena, but also provide novel resources – in a new family of models/explanations – to potentially understand them.

The position of these pioneering scientists is what I term Mutualism to distinguish it from the many other views called emergentism in the sciences and philosophy.[45] Its proponents, self-identifying as emergentists, include physicists like Laughlin, George Ellis, or Philip Anderson, chemists such as Prigogine, biologists including Denis Noble, neuroscientists like Walter Freeman, and many in the sciences of complexity or systems biology, such as Fred Boogerd, Frank Bruggeman, and Hans Westerhoff whose work we engaged in the previous section.[46]

At the heart of their positions, as I detail in Section 5.1, these scientists embrace two global ontological innovations to underpin a new guiding picture of nature. First, Mutualism accepts the Conditioned view of nature and the existence of the differential activities of parts outlined in the previous section. And second, Mutualism posits a downward whole-to-part determinative relation that I term "machresis," which I show is accompanied by a certain type of downward causation.

---

[44] See Prigogine and Stengers (1984), and Prigogine (1997), for Prigogine's interesting discussion of such ontological "myths."

[45] For surveys of the various species of "emergence," and "emergentism," see Gillett (2002b) and (2016a, ch. 5).

[46] See, for example, Anderson (1972), Boogerd et al. (2005), Ellis (2012), Freeman (2000), Laughlin (2005), Noble (2006), and Prigogine (1997).

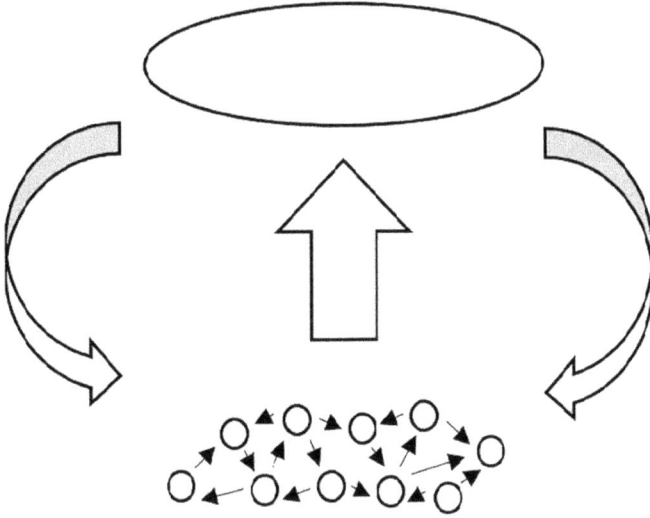

Figure 7 The complexity researcher Chris Langton's famous diagram of the scientific emergentist's innovative "Mutualist" view of nature introducing "downward" relations from composed to component entities. (Redrawn from Lewin 1992).

Given these innovations, as Figure 7 frames and as I detail in Section 5.2, under Mutualism we sometimes have compositional relations upwards from parts (and their activities and properties) *alongside* a downward determinative relation from the whole (and its activities and properties) to its parts and their differential activities (and/or powers and properties). Consequently, we have *mutual* determination between parts and whole – hence the "Mutualist" tag for the position. Crucially, as Figure 7 begins to highlight, I show the new guiding picture of Mutualism *supplements* that of Everyday Reductionism, rather than abandoning large elements of it like Fundamentalism.

Still more significantly, in Section 5.3, I further detail the new class of Mutualist models/explanations, noted in Section 4, that is offered by this position. I frame two kinds of such Mutualist models, one synchronic and the other diachronic in character, that I briefly suggest offer help in Challenging Compositional Cases. I also flag how such Mutualist models/explanations again *supplement* extant causal and compositional models, hence adding to the reper-toire of models pioneered by Everyday Reductionism, and earlier approaches, rather than abandoning them.

As I outline in Section 5.4, Mutualism also underwrites novel views of the laws and methodologies appropriate to nature once we abandon the "myth-ology" of Fundamentalism. For the new Mutualist picture implies that there are many compositional levels of both determinative parts and determinative wholes, rather than just parts and collectives of them. I sketch how Mutualism consequently opens the possibility of fundamental determinative phenomena, fundamental laws, and fundamental research at many levels of nature.

I conclude, in Section 5.5, that Mutualism is very much a Global Ontological Research Movement pursuing the Dynamic Cycle to supply new ontic models/ explanations to address ongoing problems. For the work of the section shows that Mutualism has all of features (i)–(iii) that characterized Everyday Reductionism. Ironically, given the nomenclature we have fallen into, I also suggest that Mutualist "emergentism" is indeed the real scientific heir to Everyday Reductionism in a variety of ways, whilst Fundamentalist "reduction- ism" is the outlier in scientific debates.

I briefly turn away from the scientific context, in Section 5.6, to highlight why Mutualism also has theoretical lessons to teach both philosophers and scientists. For Mutualism highlights how even when we have compositional models/ explanations we may not be able to explain everything about parts themselves solely using other parts. We thus see that we need to abandon the simple parsimony reasoning used by so many philosophers, and scientists, to get to ontologically reductive conclusions. Instead, more complicated, revised arguments need to be pursued.

I return to the sciences, in Section 5.7, where I suggest we are now in the stage of the Dynamic Cycle where we are applying rival models to specific cases in the context of investigation and preparing to use our empirical findings to evaluate these models. And I suggest that our present reduction–emergence debates again turn around issues that are empirically resolvable, hence reinfor- cing key claims of my earlier work (Gillett 2016a, ch. 10).

5.1 Two Ontological Innovations and a New Guiding Picture of Nature

We saw in Section 4 that scientific emergentists like Laughlin, or our team of systems biologists in Boogerd et al., take our evidence from Challenging Compositional Cases to show there are differential activities of individuals. The first ontological innovation of Mutualism is thus to endorse the Conditioned view of nature which we can frame in the slogan that "Parts behave differently in wholes" (Gillett 2016a).

The other ontological innovation of Mutualist emergentism is to posit that composed entities downwardly determine that their parts engage in differential activities and have differential powers. Mutualists thus champion a novel downward, synchronic determinative relation, from whole to parts (or their powers, properties, or activities), that I have elsewhere dubbed a "machretic" relation (Gillett 2016a). This is very much a categorial innovation, since we can quickly appreciate that it is a whole new category of determinative relation.

The broached downward determinative relation, from whole to parts, is not a compositional relation because a whole (and its activities and properties) cannot fill the causal role of its parts (and their activities and properties), nor hence compose them. And this downward synchronic determinative relation is also not causal, since it again has features that causal relations lack – such as being a synchronous relation, holding between entities that are in some sense the same, and involving synchronous changes in its relata, among other features. One attractive idea broached by a number of writers, since at least the theoretical biologist Howard Pattee (1970, 1973), is to take such machretic relations, at a time, to work through what are labeled the "constraints" of the relevant whole upon its parts.

It is important to note that such novel machretic relations, and their constraining influence, are plausibly always accompanied by a very particular kind of "downward causation" that is emphasized by a range of Mutualists.[47] Under Mutualism, wholes have at least two kinds of causal relations. At their own levels, wholes act "horizontally" on other wholes at the same compositional level through *thick* causal relations of activity of the whole operating within a level – thus a cell acts upon other cells. And there are plausible reasons to think that relations of activity cannot hold between entities that are compositionally related to each other or across compositional levels.[48]

In addition, however, wholes that are involved in machretic relations are also involved in "thin" causal relations of manipulability, over time, including such thin relations holding downwardly between the whole at one time and entities at other levels, including its parts, at some later time. Thick and thin causal relations differ in their features – for example, thick relations of activity require energy transfer between their relata, while thin relations of manipulability have no such requirement. Given such differences, thin manipulability relations can hold, over time, between entities at different levels. For example, through its machretic relations a whole (or its properties or activities) synchronously determines that some property of a part contributes a differential power at that time that results in a differential activity of the individual. Consequently, we will have *thin* downward causal relations of manipulability over time between the whole (and/or its relevant activity or property) and the differential activity of

---

[47] For more discussion, see Gillett (2020) and (2016a, ch. 7).
[48] See Gillett (2021, Unpublished) for arguments that such activities between parts and whole, or across compositional levels, would result in contradictions about energy.

the part and its effects at some later time.[49] And the latter are *downward* causal relations, albeit of the thin variety, since whole and parts (or their activities and properties) are at different levels.

Alongside machretic relations at a time, Mutualism thus also plausibly entails that we have a species of thin downward causation over time. We can hence see why so many Mutualists unsurprisingly frame their views around such "downward causation," though we should remember this is the thin variety and it exists alongside the machretic relations, and their constraining influence, that give rise to it.[50]

5.2 A New Guiding Picture of Nature Supplementing the Everyday Reductionist View

Considering a concrete case can help us to better appreciate Mutualism's new guiding picture of nature. For example, here is how the neuroscientist Walter Freeman applies Mutualism to a Challenging Compositional Case involving neurons in populations. Freeman tells us:

> An elementary example is the self-organization of a neural population by its component neurons. The neuropil in each area of cortex contains millions of neurons interacting by synaptic transmission. The density of action is low, diffuse and widespread. Under the impact of sensory stimulation, by the release from other parts of the brain of neuromodulatory chemicals … all the neurons come together and form a mesoscopic pattern of activity. This pattern simultaneously constrains the activities of the neurons that support it. The microscopic activity flows in one direction, upward in the hierarchy, and simultaneously the macroscopic activity flows in the other direction, downward. (Freeman 2000a, pp. 131–132)

Here we have the guiding picture of Mutualism, and its innovative ideas, applied to a concrete case. We have a whole (and its activities and properties), here a population of neurons (or tissue), upwardly composed by parts, in neurons (and their activities and properties). But at the same time this whole (and its activities and properties) also downwardly machretically determines (and "constrains") these component neurons (and/or their activities and properties), which consequently have differential behaviors and powers. We also have a range of thin downward causal relations of manipulability, not mentioned by Freeman in this passage, from the whole (and its activities and properties) to the parts (and their activities and properties) along with the effects of these differential activities.

---

[49] As I noted earlier, what I am terming thick causal relations are relations of activity. In contrast, thin causal relations are captured by manipulability or difference-making accounts that require no such relation of activity between their relata.

[50] For writers endorsing downward causation, alongside Mutualism, see the papers in Andersen et al. (2000), Noble (2008), or Ellis (2012), amongst others.

The guiding picture of Mutualism thus allows us to finally see differential activities of parts in the natural world by embracing the Conditioned view that allows the space for their existence. Furthermore, this picture also means we can finally see whole-to-part determination, and thin downward causation, whose existence Mutualism also embraces. As Freeman, Laughlin, Prigogine, and other Mutualists all emphasize, their novel ontological views revolutionize our picture of nature. But it is also important to appreciate how these categorial innovations are offered *alongside*, and *supplementing*, the commitments of successful extant approaches.

Under the Mutualist's guiding picture of nature, we need to mark how ontological elements have been *added* to the picture of Everyday Reductionism in two main ways. Mutualism has added the existence of differential activities and powers of parts at lower levels. And Mutualism has also added downward determinative relations of machresis, at a time, and consequent thin downward causal relations over time. But each of these innovations is posited *alongside* the upward compositional relations, and compositional levels of entities, endorsed by Everyday Reductionism as well as the various causal, mechanistic, kind-based, and other relations posited in our extant models.

The guiding picture of Mutualism thus plausibly supplements the guiding picture of Everyday Reductionism, rather than gutting it as Fundamentalism sought to do. As we shall see in the next subsection, the payoff with this approach is that we are again left with new kinds of models that can supplement the models in our existing coalitions of plural, but integrated, models.

5.3 Mutualist Models/Explanations as a New Resource

Following the Dynamic Cycle, Mutualism pioneers a new family of models/ explanations underpinned by its novel ontological views. We could term these "emergentist," "whole-to-part," "machretic," or "downward causal" models/ explanations, but in the previous section I simply dubbed them all Mutualist models/explanations given their connection to the wider ontological framework of Mutualism. Here I look at two kinds of such models. In Section 5.2.1, I outline a synchronic variety of mutualist model that posits machretic relations; and, in 5.2.2, I detail a diachronic kind of mutualist model built around relations of thin downward causation. I highlight how each kind of Mutualist model offers potential help with Challenging Compositional Cases.

5.3.1 Synchronic Mutualist Models

Under Mutualism one can offer a species of synchronic model that represents a synchronic downward relation at some time. Crucially, such models accept differential activities/powers of parts and represent a downward machretic relation from a whole (or its activities or properties) at some time and place as determining, and hence explaining, these differential activities/powers at the same time and place.

These synchronic Mutualist models can offer new resources to researchers at the cutting edge of science in Challenging Compositional Cases. Consider the protein which has a differential activity within the cell. In this example, scientists can now offer synchronic Mutualist models positing machretic relations between the cell (or its activities or properties) and differential behaviors/ activities of some protein. In these synchronic Mutualist models, we have potential explanations of the differential activities of proteins as resulting from the machretic determination of the relevant whole (or their activities or properties).

It is important to note that these models are selective representations, so they only represent one aspect of the state of affairs in question. And researchers will use such models alongside their existing kinds of causal, mechanistic, and compositional models which are plausibly *supplemented* by such synchronic Mutualist models and vice versa.

The latter point is unsurprising for a couple of reasons. First, we have already seen, in Section 1, how our existing coalitions of models/explanations are selective representations that supplement each other in just this way. The Mutualist is plausibly following this kind of established strategy with their new models. And second, as I noted in the previous subsection, the ontological innovations of Mutualism used in these new models *supplement* the internal ontology posited in our extant kinds of model/explanation.

5.3.2 Diachronic Mutualist Models

Under Mutualism, there are also new models that represent thin, downward causal relations of manipulability between the entities at different levels, over time, which result from the machretic relations holding at certain times. Like all models, these diachronic Mutualist models are again selective, so they only represent the relevant manipulability relations and hence make the complex state of affairs cognitively tractable. But these diachronic models are supplemented by our synchronic Mutualist models, positing synchronic machretic relations, along with our various other kinds of existing models.

For example, such a diachronic Mutualist model might represent a thin downward causal relation of manipulability between the "emergent" whole (and/or its properties or activities) at some earlier time, and the novel lower-level effects of the differential activity of this part at some later time. In this type of model, machretic relations holding at the earlier time, between the whole and part, are not represented and the represented backing relation is the thin downward causal relation.

Diachronic Mutualist models again plausibly offer new resources for researchers in Challenging Compositional Cases, since one can explain and understand the differential activity of a part, and/or its effects, using the represented downward causal relation. For example, one could explain the differential activity of a protein, and its effects, at some time by pointing to the presence of the protein in the cell (with its specific activities and/or properties) at some earlier time. Although not represented in the model, the cell (and/or its activities or properties) machretically determines that the protein has the differential power and activity. Once again, such diachronic models *supplement* not only our synchronic Mutualist models, but also our existing kinds of causal, mechanistic, and compositional models.

There are still other kinds of models underwritten by Mutualism, but the two kinds I have begun to discuss illuminate some of its new explanatory resources. For we can thus see that Mutualism does indeed underwrite a new family of models offering help in Challenging Compositional Cases and beyond. And we have also found that Mutualism mirrors the strategy of Everyday Reductionism by offering ontological innovations that underpin models that supplement, and hence integrate with, the internal ontology of the existing kinds of models in our coalitions.

5.4 Mutualism, Fundamental Laws, and New Methodologies: Beyond the Fundamentalist Picture

Let us briefly turn to what Mutualism says about laws and appropriate methodologies. Unsurprisingly, as I began to sketch in Section 4, the endorsement of the Conditioned view of nature entails the existence of new fundamental laws, and methodologies that clash with the seductive Fundamentalist "mythology."

The Mutualist, and condensed matter physicist, Philip Anderson famously flags how accepting compositional models/explanations may all too easily suck us into Fundamentalist assumptions. In his pioneering paper "More Is Different," Anderson cautions us that once we accept Everyday Reductionism and its compositional models:

> It seems *inevitable* ... [to accept] what appears at first sight to be an obvious corollary of [Everyday] reductionism: that if everything obeys the same fundamental laws, then the only scientists who are studying anything really fundamental are those working on those laws ... This point of view ... it is the main purpose of this article to oppose. (Anderson 1972, p. 393)

Here we find a warning that the claims about the compositional structure of nature, and its associated laws, illuminated by Everyday Reductionism can all too easily be assumed to lead to the Fundamentalist's austere – and radical – views of the structure of nature, fundamental laws, and research.

But Anderson highlights how the alternative guiding picture of Mutualism, compatible with the approach of Everyday Reductionism and the truth of its compositional models/explanations, leads to a different picture of such laws and research. Focusing on our more detailed empirical findings about the behavior of the parts in wholes, in examples like Challenging Compositional Cases, Anderson contends that:

> The behavior of large and complex aggregations of elementary particles, it turns out, is not to be understood in terms of a simple extrapolation of the properties of a few particles. Instead, at each level of complexity entirely new properties appear, and the understanding of the new behaviors requires research which I think is as fundamental in its nature as any other. (Anderson 1972, p. 393)

This is an elegant statement of what I have termed the Conditioned view of nature and its Mutualist application to cases involving parts. And we can see that the result is a very different account of both fundamental laws and research methodologies.

Let us briefly unpack these implications starting with the nature of laws under Mutualism. For example, Laughlin tells us:

> From the [Fundamentalist] reductionist standpoint, physical law is the motivating impulse of the universe. It does not come from anywhere and implies everything. From the emergentist perspective, physical law is a rule of collective behavior, it is a consequence of more primitive rules of behavior underneath (although it need not have been), and it gives one predictive power over a limited range of circumstances. Outside this range, it becomes irrelevant, supplanted by other rules that are either its children or its parent in the hierarchy of descent. (Laughlin 2005, p. 80)

Here we find Laughlin pointing to the complex new array of fundamental laws implied by the commitments of Mutualism, given the existence of differential activities of parts and their machretic determination by certain wholes. For example, some laws applying to parts in the simplest systems may be supplemented, under the condition of being in certain wholes, by further fundamental laws about the differential activities of parts machretically determined by the relevant wholes. Or, as Laughlin outlines, the laws about parts in simple systems may sometimes be completely supplanted, under certain conditions, by the laws about the machretic determination of such parts and their differential activities by the relevant wholes. ·

The nature of such fundamental laws under Mutualism, and the various options for them, deserves much more discussion than I can give to them here. However, I have explored these issues, and some of the options, in more detail elsewhere (Gillett 2016a, ch. 7). For our present purposes, it is more interesting to note how Mutualism also leads to novel methodologies and other stark differences with the claims of Fundamentalism.

Under Fundamentalism, fundamental physics has a monopoly on fundamental phenomena, research, and laws. But, as Anderson and Laughlin (Laughlin 2005, pp. 5–8) both emphasize, under the guiding picture of Mutualism *many* entities are taken to be determinative as well as parts, so *many* sciences, studying nature at *many* levels, are required to investigate fundamental phenomena and fundamental laws of the kind just outlined. Mutualism thus implies a starkly different picture than that of Fundamentalism about the appropriate methodology for fundamental research, including our allocation of research funding. And Mutualism has a range of other methodological implications, both great and small, that writers like Sandra Mitchell have carefully illuminated (Mitchell 2009).

5.5 Contemporary Scientific Emergentism as a Global Ontological Research Movement

The work of this section supports the conclusion that Mutualism is a full-blown Global Ontological Research Movement in the sciences. For we have now found that Mutualism mirrors all the characteristics (i)–(iii) of Everyday Reductionism. To summarize, previous subsections have highlighted how:

(i) Mutualism has global/categorial ontological innovations, in its endorsement of (a) the Conditioned view and the existence of differential activities, and (b) acceptance of machretic determination of the differential activities of parts by wholes, and a new guiding picture of nature built around these innovations;

(ii) Mutualism has a novel family of Mutualist models/explanations positing differential activities/powers and underwritten by machretic relations, and/ or accompanying thin downward causal relations, resulting from its ontological innovations in (i);

And:

(iii) Mutualism has new methodologies, driven by its commitments in (i) and (ii), including the new strategies for fundamental research, and its funding, among others.

Mutualism is thus plausibly like Everyday Reductionism in being a Global Ontological Research Movement in the sciences. And, beyond mirroring its broad characteristics, we have also found further, deeper connections between Mutualism and Everyday Reductionism.

For example, we saw that Mutualism presses ontological innovations supplementing the guiding picture of nature of Everyday Reductionism. And hence Mutualism provides a new family of ontic models/explanations that consequently supplements those models pioneered by Everyday Reductionism. Mutualism thus also continues the same underlying general strategy of Everyday Reductionism: namely, that of providing new models supplementing existing models, here the models of Everyday Reductionism itself, to understand still further aspects of the relevant phenomena in nature such as the differential activities of parts in Challenging Compositional Cases.

The resulting contrast between Fundamentalism and Mutualism is striking. For we saw, in Section 3, that Fundamentalism is not focused on solving ongoing scientific problems and provides no new ontic models/explanations to do so, but is instead focused on theoretical arguments driven by parsimony considerations alone – hence mirroring the practices of the exogenous metaphysics of philosophy. Fundamentalism is thus indeed the outlier position within the sciences. And we can hence also support our diagnosis of why Fundamentalism may have been less popular among working scientists than emergentist approaches. For Mutualism does indeed supply researchers with new models/explanations to use in their work, while the Fundamentalist provides no resources of this kind.

These latter points thus bring out a still broader way in which Mutualism is like Everyday Reductionism: Both are pursuing the Dynamic Cycle by developing new ontological concepts and hence new models to address ongoing scientific problems. Though our terminology might suggest otherwise, the heir to Everyday Reductionism in the sciences therefore more plausibly appears to be Mutualist emergentism rather than Fundamentalist reductionism.

5.6 The Lessons of Mutualism about Simple Parsimony Reasoning

Let me briefly digress from the scientific context to note an important wider implication of appreciating Mutualism and its novel ontological framework. As we have seen, philosophers and scientists have each sought to use simple versions of ontological parsimony arguments about compositional relations, and/or models/explanations, to conclude that we should never accept both whole and parts. However, once we appreciate Mutualism, we can see that such simple parsimony arguments are plausibly either invalid or unsound depending upon their precise formulation. This is important, since it shows that a whole class of arguments commonly used by both scientists and philosophers to press Fundamentalism, and other ontologically reductionist positions, need to be abandoned.

Let me briefly highlight the underlying point using my formulation of such simple parsimony reasoning in the Argument from Composition, but the broader point generalizes and can plausibly be adapted to apply to other formulations of such reasoning and to related arguments such as Kim's, noted in Section 3, focused on causal relations.

Recall that the crucial subconclusion of the Argument from Composition, and related parsimony reasoning, is that in cases of compositional models/explanations one can use parts alone to account for, and explain, everything about both the whole and the parts. However, when Mutualism is true, we have differential behaviors that are machretically determined by a whole, although the whole (and its activities or properties) are the subjects of successful compositional models/explanations and are fully composed. In such a situation, despite having successful compositional models/explanations, we still *cannot* account for all the behaviors of individuals solely using parts or their activities/properties alone – for the differential behaviors of parts have not been explained.

The type of case framed by Mutualism thus shows that the truth of the crucial subconclusion of the Argument from Composition does not follow from the truth of the premise or premises. It is true that we have successful compositional models/explanations, but the subconclusion that we can explain everything with parts alone is still false – so we can see that the Argument from Composition is invalid. Similar points suffice to establish either the invalidity, or the unsound nature of the premises, in other formulations of such simple ontological parsimony arguments.

If this diagnosis is correct, then for centuries a flawed kind of theoretical argument has wrongly been used to dismiss natural phenomena that did not fit the Fundamentalist's, and other ontological reductionists', Simple view of nature whose truth is an empirical matter. We therefore finally need to put these flawed theoretical arguments to one side.

It is important to note that one can construct more adequate, but more complicated, parsimony arguments with further premises not used in simple versions (see Gillett 2016a, ch. 8). But these more complicated arguments involve premises whose truth depends upon just the contested empirical issues around which our present reduction–emergence debates can be seen to turn, such as the Simpler view.[51] Hence theoretical arguments, whether the flawed

---

[51] For example, one can add as a premise what philosophers term the "Completeness of Physics" to the simple parsimony reasoning. However, the Completeness of Physics, like a Final Theory, plausibly has the truth of the Simple view of nature as a precondition of its truth.

simple versions or the more complex varieties that should replace them, do not by themselves resolve the debates as many philosophers, and various Fundamentalist scientists, have claimed. Instead, it appears that empirical – and hence scientific – inquiry has a crucial role. Let us therefore return to the present situation in the scientific discussions and see whether the deeper issues are indeed plausibly empirically resolvable.

5.7 The Present Stage of our Reduction–Emergence Debate

At what stage are we now in the second cycle of scientific reduction–emergence debates? Researchers are presently exploring in various scientific examples, in the context of investigation, what appear to be Mutualist models, and also rival models such as the Causally Conditioned ones sketched in the previous section. I therefore suggest that we are now plausibly at the stage in the Dynamic Cycle where we are gathering empirical evidence relevant to assessing these models. Such work is the prelude to the next stage of the Dynamic Cycle, where scientists can assess, in each scientific case, which of these models, and their ontological concepts, should be retained, revised, supplemented, or replaced.

Such evaluations of competing models do not turn around whether they are popular among working scientists – so the relative popularity of Mutualism, or related views, that I noted earlier is not relevant. Rather, the central question in this stage of the Dynamic Cycle is which of our models, given our evidence, has an internal ontology best reflecting what our evidence suggests is the ultimate ontology in each example. Whether Mutualist models, or Causally Conditioned ones, or some other variety, most successfully describe and explain the behavior of parts in Challenging Compositional Cases is thus plausibly a central question on a case-by-case basis in such evaluations.

It also consequently appears that, like the key metaphysical issues of the first set of debates, the deeper questions of our present reduction–emergence debates are empirically resolvable in a broad sense. The relative success of the models they support, in specific scientific examples, along with other scientific evidence, can potentially resolve the question of whether the Conditioned or Simple views of nature are correct. Or whether differential activities of parts exist and/or are always determined by other individuals at their own levels (as Causally Conditioned models posit) or are sometimes determined by the whole (or its activities or properties) that they compose (as Mutualist models frame).

Like the first set of scientific debates, as we saw in the previous subsection, these latter questions are not simply resolvable by theoretical arguments alone, contrary to the contention of Fundamentalist scientists and too many onto-logical reductionists in philosophy. Rather, if we are to have answers to these questions, then we must patiently wait for the underlying scientific processes involved in the Dynamic Cycle, namely those of applying models, securing empirical findings, comparatively evaluating models, and so on, to slowly, and carefully, play out.

5.8 Conclusion: Appreciating the Wider Import of Endogenous Metaphysics and the Dynamic Cycle

Philosophers, whether metaphysicians or philosophers of science, often have difficulty seeing that there might be other kinds of metaphysics pursued outside of philosophy. One referee for this Element, commenting on an admittedly rough draft, claimed to find endogenous metaphysics "mysterious." One often also finds philosophers puzzled by the idea that the sciences could resolve "metaphysical" questions. And some philosophers of science are still deeply resistant to the idea that "metaphysics" might ever be productive for the sciences. As well as illuminating the positions in our two sets of scientific reduction–emergence debates, I hope I have now made a case speaking to each of these philosophical concerns.

Once one appreciates that scientific explanations are often models driven by their representations of entities in nature, underwritten by the ontological concepts of the scientists offering these models, then it would in fact be a mystery how science could proceed *without* endogenous metaphysics. Furthermore, although endogenous metaphysics is usually pursued locally, and incrementally, we have seen how in the first set of reduction–emergence debates a Global Ontological Research Movement pioneered a categorial ontological innovation to underwrite a new family of compositional models to help with ongoing scientific problems. In fact, we saw how the resulting Everyday Reductionist movement, across the twentieth century, *resolved* the question of whether all working individuals in nature (and their activities and working properties) are composed by the entities of physics – hence answering a metaphysical question. Once we appreciate the existence of compositional models/explanations and the Dynamic Cycle, we thus also see how researchers pursuing endogenous metaphysics were spectacularly successful in twentieth-century science.

I should emphasize in concluding that I have not sought to illuminate endogenous metaphysics to denigrate any other kinds of metaphysics that philosophers or others engage in. Like most intellectual areas, metaphysics is presently pursued in plural ways by analytic philosophers and others. As I hope this Element begins to make clear, I am a thorough-going *pluralist* – I think there are many questions about reality, each requiring different answers in distinct kinds of selective representation, usually with a proprietary internal ontology, and often offered by distinct areas of inquiry using divergent methods. Unsurprisingly, I therefore believe there are ontological questions that are *not* plausibly answerable by the sciences using its endogenous metaphysics and the Dynamic Cycle, and for which other kinds of metaphysics are often better suited.

Let me also briefly note that there are various questions raised by my work here that I have not explored at all given the tight space limitations. For example, have there been other Global Ontological Research Movements in the sciences beyond reduction–emergence debates?[52] What were the onto-logical innovations of these movements and how successful were they? Such questions, and others, all deserve attention.

To take another instance, my work also raises the issue of what kind of integration we find between the models in our coalitions in the context of understanding. The story of this Element, across the two sets of reduction–emergence debates, suggests scientists in physiology, cell biology, and molecu-lar biology successfully developed new kinds of models/explanations to *sup-plement* the models in their existing coalitions.[53] But whether this situation holds more broadly for these sciences, or other areas of the sciences, and to what degree, also needs further exploration.

To conclude, I hope you agree that revisiting, and rethinking, scientific reduction–emergence debates has been rewarding. For it allowed us to appreci-ate the importance of the endogenous metaphysics pursued in foundational scientific practices such as the Dynamic Cycle of ontic concept/model creation. Looking narrowly, we were also able to finally appreciate the existence, and achievements, of the Everyday Reductionist movement in twentieth-century science and to understand key questions, and ongoing responses, in contempor-ary movements at the cutting edge of contemporary science. Looking more broadly, as we have just seen, an exciting array of new questions come into view about how, when, and why scientists develop new ontological concepts, and associated families of model/explanation, and whether or how they extend our understanding of nature.

---

[52] Thanks to Alex Carruth for pressing this important issue.
[53] See Gillett (Unpublished, ch. 7), for a more detailed account of such integration.

References

Aizawa, K. 2007. "The Biochemistry of Memory Consolidation." *Synthese*, 155, pp. 65–98.

Aizawa, K. and Gillett, C. 2014. "Realization, Reduction and Emergence." In N. Levy (ed.), *Handbook of Neuroethics*. Dordrecht: Springer, pp. 49–62.

Aizawa, K. and Gillett, C. (eds.) 2016a. *Scientific Composition and Metaphysical Grounding*. New York, NY: Palgrave Macmillan. DOI: https://doi.org/10.1057/978-1-137-56216-6_1.

Aizawa, K. and Gillett, C. 2016b. "Vertical Relations in Science, Philosophy and the World: Understanding the New Debates over Verticality." In K. Aizawa and C. Gillett (eds.), *Scientific Composition and Metaphysical Ground*, pp. 1–38. New York, NY: Palgrave Macmillan.

Aizawa, K. and Gillett, C. 2019. "Defending Pluralism about Compositional Explanations." *Studies in the History and Philosophy of Science, Part C*, 78, 101202.

Alexander, S. 1920. *Space, Time and Deity*. Two volumes. The Gifford Lectures 1916–18. Toronto: Macmillan.

Andersen, H. and Mitchell, S. (eds.) 2023. *The Pragmatist Challenge: Pragmatist Metaphysics for Philosophy of Science*. New York, NY: Oxford University Press.

Andersen, P., Christiansen, P., Emmeche, C., and Finnemann, N. (eds.) 2000. *Downward Causation: Minds, Bodies and Matter*. Aarhus: Aarhus University Press.

Anderson, P. 1972. "More Is Different: Broken Symmetry and the Nature of the Hierarchical Structure of Science." *Science*, 177, pp. 393–96.

Auyang, S. 1998. *Foundations of Complex-Systems Theories in Economics, Evolutionary Biology, and Statistical Physics*. Cambridge: Cambridge University Press.

Bechtel, W. 2005. "Explanation: A Mechanistic Alternative." *Studies in the History and Philosophy of Biology and Biomedical Sciences*, 36, pp. 421–41.

Bedau, M. 1997. "Weak Emergence." *Philosophical Perspective*, 11, pp. 375–99.

Betts, J. G., Young, K., Wise, A. et al. (eds.) 2013. *Anatomy and Physiology*. BC Campus. Available at https://openstax.org/books/anatomy-and-physiology/pages/1-introduction.

Bickle, J., Craver, C., and Barwich, A. 2022. *The Tools of Neuroscience Experiment*. New York, NY: Routledge.

Bishop, R., Silberstein, M., and Pexton, M. 2022. *Emergence in Context.* New York, NY: Oxford University Press.

Bollhagen, A. 2021. "The Inchworm Episode: Reconstituting the Phenomenon of Kinesin Motility." *European Journal of Philosophy of Science*, 11, pp. 1–25.

Boogerd, F., Bruggeman, F., Richardson, R., Stephan, A., and Westerhoff, H. 2005. "Emergence and Its Place in Nature: A Case Study of Biochemical Networks." *Synthese*, 145, pp. 131–64.

Brigandt, I. 2011. "Explanation in Biology: Reduction, Pluralism and Explanatory Aims." *Science and Education*, 22, pp. 69–91.

Brigandt, I. and Love, A. 2023. "Reductionism in Biology." In E. Zalta and U. Nodelman (eds.), *The Stanford Encyclopedia of Philosophy.* https://plato .stanford.edu/archives/sum2023/entries/reduction-biology/.

Broad, C. D. 1925. *The Mind and Its Place in Nature.* London: Routledge Kegan Paul.

Cartwright, N. 1994. "Fundamentalism vs. the Patchwork of Laws." *Proceedings of the Aristotelian Society*, 103, pp. 279–92.

Couzin, I. and Krause, J. 2003. "Self-Organization and Collective Behavior in Vertebrates." *Advances in the Studies of Behavior*, 33, pp. 1–75.

Craver, C. 2007. *Explaining the Brain.* New York, NY: Oxford University Press.

Crick, F. 1966. *Of Mice and Molecules.* Seattle, WA: University of Washington.

Crook, S. and Gillett, C. 2001. "Why Physics Alone Cannot Define the 'Physical'." *Canadian Journal of Philosophy*, 31, pp. 333–60.

Dawkins, R. 1982. *The Extended Phenotype.* San Francisco, CA: W. H. Freeman.

Dawkins, R. 1987. *The Blind Watchmaker.* New York, NY: Norton.

Dennett, D. 1978. *Brainstorms.* Montgomery, VT: Bradford Books.

Dennett, D. 1991. "Real Patterns." *Journal of Philosophy*, 88, pp. 27–51.

Dennett, D. 1996. *Darwin's Dangerous Idea.* New York, NY: Simon and Schuster.

Driesch, H. 1929. *The Science and Philosophy of the Organism.* 2nd ed. London: Black.

Dupre, J. 2021. *The Metaphysics of Biology.* New York, NY: Cambridge University Press.

Eddington, A. S. 1928. *The Nature of the Physical World.* New York, NY: Macmillan.

Ellis, G. 2012. "Top-Down Causation and Emergence." *Interface Focus*, 2, pp. 126–40.

Elpidorou, A. and Dove, G. 2018. *Consciousness and Physicalism: A Defense of a Research Program.* New York, NY: Routledge.

Fodor, J. 1968. *Psychological Explanation.* New York, NY: Random House.

Fodor, J. 1974. "Special Sciences: Or, the Disunity of Science as a Working Hypothesis." *Synthese*, 28, pp. 97–115.

Freeman, W. 2000a. *How Brains Make Up Their Minds*. New York, NY: Columbia University Press.

Freeman, W. 2000b. *Neurodynamics: An Exploration of Mesoscopic Brain Dynamics*. London: Springer.

Garfinkel, A. 1987. "The Slime Mold *Dictyostelium* as a Model of Self-Organization in Social Systems." In F. Yates, A. Garfinkel, D. Walter, and G. Yates (eds.), *Self-Organizing Systems*, pp. 181–212. New York, NY: Plenum Press.

Giere, R. 2010. *Scientific Perspectivism*. Chicago, IL: University of Chicago Press.

Gillett, C. 2002a. "The Varieties of Emergence: Their Purposes, Obligations and Importance." *Grazer Philosophische Studien*, 65, pp. 89–115.

Gillett, C. 2002b. "The Dimensions of Realization: A Critique of the Standard View." *Analysis*, 62, pp. 316–23.

Gillett, C. 2006. "Samuel Alexander's Emergentism: Or, Higher Causation for Physicalists." *Synthese*, 153, pp. 261–96.

Gillett, C. 2007a. "Understanding the New Reductionism: The Metaphysics of Science and Compositional Reduction." *The Journal of Philosophy*, 104(4), pp. 193–216.

Gillett, C. 2007b. "Hyper-Extending the Mind? Setting Boundaries in the Special Sciences." *Philosophical Topics*, 351, pp. 161–88.

Gillett, C. 2011. "On the Implications of Scientific Composition and Completeness." In A. Corradini and T. O'Connor (eds.), *Emergence in Science and Philosophy*, pp. 25–45. New York, NY: Routledge.

Gillett, C. 2013. "Constitution, and Multiple Constitution, in the Sciences: Using the Neuron as a Guide." *Minds and Machines*, 23, pp. 309–37.

Gillett, C. 2016a. *Reduction and Emergence in Science and Philosophy*. New York, NY: Cambridge University Press.

Gillett, C. 2016b. "The Metaphysics of Nature, Science, and the Rules of Engagement." In K. Aizawa and C. Gillett (eds.), *Scientific Composition and Metaphysical Ground*, pp. 205–47. New York, NY: Palgrave Macmillan.

Gillett, C. 2020. "Why Constitutive Mechanistic Explanation Cannot Be Causal." *American Philosophical Quarterly*, 57, pp. 31–50.

Gillett, C. 2021. "Using Compositional Explanations to Understand Compositional Levels." In D. Brooks, J. DeFrisco, and W. Wimsatt (eds.), *Levels of Organization in the Biological Sciences*, pp. 233–60. Cambridge, MA: MIT Press.

Gillett, C. 2022. "Engaging the Plural Parts of Science." *Journal of Consciousness Studies*, 29, pp. 195–217.

Gillett, C. Unpublished. *Understanding the Human Body: The Parts of Science, Compositional Models and Integrative Pluralism*.

Glennan, S. 2017. *The New Mechanical Philosophy*. New York, NY: Oxford University Press.

Glennan, S. 2020. "Corporeal Composition." *Synthese*, 198, pp. 11439–62.

Guay, A. and Sartenaer, O. 2016. "A New Look at Emergence: Or When After is Different." *European Journal of Philosophy of Science*, 6, pp. 297–322.

Haraway, D. 1976. *Crystals, Fabrics and Fields*. New Haven, CT: Yale University Press.

Heil, J. 2003. *From an Ontological Point of View*. New York, NY: Oxford University Press.

Hempel, C. 1965. *Aspects of Scientific Explanation and Other Essays in the Philosophy of Science*. New York, NY: Free Press.

Hendry, R. 2010. "Emergence vs. Reduction in Chemistry." In C. MacDonald and G. MacDonald (eds.), *Emergence in Mind*, pp. 205–21. New York, NY: Oxford University Press.

Humphreys, P. 2016. *Emergence: A Philosophical Account*. New York, NY: Oxford University Press.

Illari, P. and Williamson, J. 2012. "What Is a Mechanism?" *European Journal for Philosophy of Science*, 2(1), pp. 119–35.

Juarrero, A. 1999. *Dynamics in Action: Intentional Behavior as a Complex System*. Cambridge, MA: MIT Press.

Kaiser, M. 2018. "Individuating Part-Whole Relations in the Biological World." In O. Bueno, R-L. Chen, and M. Fagan (eds.), *Individuation across Experimental and Theoretical Sciences*, pp. 63–88. New York, NY: Oxford University Press.

Kim, J. 1993a. *Supervenience and Mind*. Cambridge: Cambridge University Press.

Kim J. 1993b. "The Nonreductionist's Troubles with Mental Causation." In J. Kim (1993a).

Kitcher, P. 1984. "1953 and All That: A Tale of Two Sciences." *Philosophical Review*, 93, pp. 335–73.

Ladyman, J. and Ross, D. 2007. *Everything Must Go*. New York, NY: Oxford University Press.

Laudan, L. 1977. *Progress and its Problems*. Berkeley, CA: University of California Press.

Laughlin, R. 2005. *A Different Universe: Reinventing Physics from the Bottom Down*. New York, NY: Basic Books.

Lewin, R. 1992. *Complexity: Life at the Edge of Chaos*. New York, NY: Macmillan.

Love, A. 2012. "Hierarchy, Causation and Explanation." *Interface Focus*, 2, pp. 115–25.

Love, A. and Huttemann, A. 2011. "Comparing Part-Whole Explanations in Biology and Physics." In D. Dieks, W. J. Gonzalez, S. Hartmann, et al. (eds.), *Explanation, Prediction, and Confirmation*, pp. 183–200. Berlin: Springer.

Machamer, P., Darden, L., and Craver, C. 2000. "Thinking about Mechanisms." *Philosophy of Science*, 67, pp. 1–25.

Mayr, E. 1988. "The Limits of Reductionism." *Nature*, 33, p. 475.

McLaughlin, B. 1992. "The Rise and Fall of British Emergentism." In A. Beckermann, H. Flohr, and J. Kim (eds.). *Emergence or Reduction?*, pp. 49–93. New York, NY: de Gruyter.

Merricks, T. 2001. *Objects and Persons*. New York, NY: Oxford University Press.

Mitchell, S. 2003. *Biological Complexity and Integrative Pluralism*. Cambridge: Cambridge University Press.

Mitchell, S. 2009. *Unsimple Truths: Science, Complexity and Policy*. Chicago, IL: University of Chicago Press.

Mitchell, S. 2012. "Emergence: Logical, Functional and Dynamical." *Synthese*, 185, pp. 171–86.

Moreno, A. and Mossio, M. 2015. *Biological Autonomy*. New York, NY: Springer.

Nagel, E. 1961. *The Structure of Science*. New York, NY: Harcourt Brace.

Needham, J. 1936. *Order and Life*. New Haven, CT: Yale University Press.

Nicolis, G. and Prigogine, I. 1989. *Exploring Complexity*. New York, NY: Freeman.

Noble, D. 2008. *The Music of Life: Biology beyond Genes*. New York, NY: Oxford University Press.

Novikoff, A. 1945. "The Concept of Integrative Levels and Biology." *Science*, 101, pp. 209–215.

Pattee, H. 1970. "The Problem of Biological Hierarchy." In C. H. Waddington, *Towards a Theoretical Biology*, 3, pp. 117–136. Chicago, IL: Aldine Publishing.

Pattee, H. 1973. "The Physical Basis and Origin of Hierarchical Control." In H. Pattee, *Hierarchy Theory*, pp. 91–110. New York, NY: George Braziller.

Pereboom, D. 2002. "Robust Nonreductive Materialism." *Journal of Philosophy*, 99, pp. 499–531.

Pereboom, D. 2011. *Consciousness and the Prospects of Physicalism*. Oxford: Oxford University Press.

Prigogine, I. 1968. *Introduction to Thermodynamics of Irreversible Processes.* New York, NY: Wiley.

Prigogine, I. 1997. *End of Certainty.* New York, NY: The Free Press.

Prigogine, I. and Stengers, I. 1984. *Order out of Chaos: Man's New Dialogue with Nature.* New York, NY: Bantam Books.

Rosenberg, A. 2006. *Darwinian Reductionism.* Chicago, IL: Chicago University Press.

Salmon, W. 1989. *Four Decades of Scientific Explanation.* Minneapolis, MN: University of Minnesota Press.

Santos, G. 2015. "Ontological Emergence: How Is It Possible? Towards a New Relational Ontology." *Foundations of Science*, 20, pp. 429–446.

Scott, A. 2007. *The Non-Linear Universe: Chaos, Emergence, Life.* New York, NY: Springer.

Shoemaker, S. 2007. *Physical Realization.* New York, NY: Oxford University Press.

Siderits, M. 2007. *Buddhism as Philosophy: An Introduction.* Indianapolis, IN: Hackett.

Sullivan, J. 2017. "Coordinated Pluralism as a Means to Facilitate Integrative Taxonomies of Cognition." *Philosophical Explorations*, 20, pp. 129–145.

Tahko, T. 2021. *Unity of Science.* New York, NY: Cambridge University Press.

Thalos, M. 2013. *Without Hierarchy: The Scale Freedom of the Universe.* Oxford: Oxford University Press.

Van Inwagen, P. 1990. *Material Beings.* Ithaca, NY: Cornell University Press.

Weinberg, S. 1992. *Dreams of a Final Theory.* New York, NY: Random House.

Weinberg, S., 2001. *Facing Up: Science and Its Cultural Adversaries.* Cambridge, MA: Harvard University Press.

Wilson, E. 1998. *Consilience: The Unity of Knowledge.* New York, NY: Knopf.

Wilson, E. and Holldobler, B. 1988: "Dense Heterarchies and Mass Communication as the Basis of Organization in Ant Colonies." *Trends in Ecology and Evolution*, 3, pp. 65–84.

Wimsatt, W. 1976. "Reductionism, Levels of Organization and the Mind-Body Problem." In *Consciousness and the Brain*, edited by G. Globus, G. Maxwell, and I. Savodnik, pp. 205–267. New York, NY: Plenum Press.

Wimsatt, W. 2007. *Re-engineering Philosophy for Limited Beings.* Cambridge, MA: Harvard University Press.

Winning, J. and Bechtel, W. 2018. "Rethinking Causality in Biological and Neural Mechanisms." *Minds and Machines*, 28, pp. 287–310.

Woodger, J. H. 1929. *Biological Principles.* New York, NY: Harcourt, Brace.

Wright, C. 2012. "Mechanistic Explanation without the Ontic Conception." *European Journal of Philosophy of Science*, 2, pp. 375–394.

Acknowledgments

I am grateful to a number of people for discussing the ideas of this Element including Ken Aizawa, David Barack, John Bickle, Alex Carruth, Sandy Mitchell, and Alok Srivastava. Thanks to Shaun Nichols for encouraging me to write an introductory book and to Brian McLaughlin for long ago getting me interested in these rich debates. I am also grateful to two anonymous referees for their helpful comments on an overly rough draft. Special thanks to my editor Tuomas Tahko for inviting me to contribute this Element and for his help in bringing it into existence.

Cambridge Elements ≡

Metaphysics

Tuomas E. Tahko

University of Bristol
Tuomas E. Tahko is Professor of Metaphysics of Science at the University of Bristol, UK. Tahko specialises in contemporary analytic metaphysics, with an emphasis on methodological and epistemic issues: 'meta-metaphysics'. He also works at the interface of metaphysics and philosophy of science: 'metaphysics of science'. Tahko is the author of *Unity of Science* (Cambridge University Press, 2021, *Elements in Philosophy of Science*), *An Introduction to Metametaphysics* (Cambridge University Press, 2015), and editor of *Contemporary Aristotelian Metaphysics* (Cambridge University Press, 2012).

About the Series

This highly accessible series of Elements provides brief but comprehensive introductions to the most central topics in metaphysics. Many of the Elements also go into considerable depth, so the series will appeal to both students and academics. Some Elements bridge the gaps between metaphysics, philosophy of science, and epistemology.

Cambridge Elements ☰

Metaphysics

Elements in the Series

For EU product safety concerns, contact us at Calle de José Abascal, 56–1°, 28003 Madrid, Spain or eugpsr@cambridge.org.

www.ingramcontent.com/pod-product-compliance
Ingram Content Group UK Ltd.
Pitfield, Milton Keynes, MK11 3LW, UK
UKHW021512110325
456102UK00006B/56